Politics and Practice in Economic Geography

Politics and Practice in Economic Geography

Edited by
Adam Tickell
Eric Sheppard
Jamie Peck
Trevor Barnes

SAGE Publications
Los Angeles · London · New Delhi · Singapore

First published 2007

SAGE Publications Ltd
1 Oliver's Yard
55 City Road
London EC1Y 1SP

SAGE Publications Inc.
2455 Teller Road
Thousand Oaks, California 91320

SAGE Publications India Pvt Ltd
B 1/I 1 Mohan Cooperative Industrial Area
Mathura Road
New Delhi 110 044

SAGE Publications Asia–Pacific Pte Ltd
33 Pekin Street #02-01
Far East Square
Singapore 048763

Library of Congress Control Number: 2007922922

British Library Cataloguing in Publication data

A catalogue record for this book is available
from the British Library

ISBN 978-1-4129-0785-9
ISBN 978-1-4129-0786-6 (pbk)

Typeset by C&M Digitals (P) Ltd., Chennai, India
Printed in India at Replika Press Pvt. Ltd.
Printed on paper from sustainable resources

CONTENTS

LIST OF FIGURES AND TABLES

LIST OF CONTRIBUTORS

Trevor Barnes, Department of Geography, University of British Columbia

Judith Carney, Department of Geography, University of California, Los Angeles

Gordon L. Clark, School of Geography, University of Oxford

Altha J. Cravey, Department of Geography, University of North Carolina at Chapel Hill

Elizabeth C. Dunn, Geography and International Affairs, University of Colorado at Boulder

Susan Geiger retired from the University of Minnesota in 1999 and passed away after a 16-month battle with leukaemia in April 2001

J.K. Gibson-Graham, Department of Geosciences, University of Massachusetts at Amherst/Department of Human Geography, Australia National University

Vinay K. Gidwani, Department of Geography, University of Minnesota

Amy Glasmeier, Department of Geography, Pennsylvania State University

Jim Glassman, Department of Geography, University of British Columbia

Caleb Johnston, Department of Geography, University of British Columbia

Philip F. Kelly, Department of Geography, York University

Mei-Po Kwan, Department of Geography, Ohio State University

Linda McDowell, School of Geography, University of Oxford

Doreen Massey, Faculty of Social Sciences – Geography, The Open University

Richard Meegan, European Institute for Urban Affairs, Liverpool John Moores University

Alison Mountz, Department of Geography, Syracuse University

Richa Nagar, Department of Gender, Women, and Sexuality Studies, University of Minnesota

Kris Olds, Department of Geography, University of Wisconsin-Madison

Phillip O'Neill, Urban Research Centre, University of Western Sydney

Jamie Peck, Department of Geography, University of Wisconsin-Madison

John Pickles, Department of Geography, University of North Carolina at Chapel Hill

Paul Plummer, Department of Geography, University of Calgary

Geraldine Pratt, Department of Geography, University of British Columbia

David L. Rigby, Department of Geography, University of California, Los Angeles

Paul Robbins, Department of Geography and Regional Development, University of Arizona

Andrew Sayer, Department of Sociology, University of Lancaster

Erica Schoenberger, Department of Geography and Environmental Engineering, The Johns Hopkins University

Eric Sheppard, Department of Geography, University of Minnesota

Adrian Smith, Department of Geography, Queen Mary, University of London

Adam Tickell, Faculty of History and Social Science, Royal Holloway, University of London

Jane Wills, Department of Geography, Queen Mary, University of London

Henry Wai-chung Yeung, Department of Geography, National University of Singapore

FOREWORD

Doreen Massey and Richard Meegan

It came as something of a jolt to realize that it is over twenty years now since we convened the experimental debate between industrial geographers that was to result in *Politics and Method*. Times have changed. Not only have the debates within economic geography shifted radically, but the wider socio-political context has also been transformed. When we gathered for our discussions at the Open University, in 1983, while it was clear that the old model of accumulation was on the rocks, it was not clear what was to emerge in its place. The real economic geography of the UK was embattled, partly in defence of the past, partly around what might be the future. The steelworkers' strike of 1980, the ongoing battles over inner cities, the too-late attempt by a previously complacent British capitalism to restructure (the lateness in consequence leading to far more devastation than it might have) … deindustrialization everywhere. All this formed an important context for our gathering.

Our aim was to explore both how industrial geographers might analyze such a situation and, of equal importance, the kinds of policy/political stances they should take. But we were caught up in changing times in another way too. There was fierce debate, across the social sciences more widely, about method, and our aim was also to address that. We wanted genuinely to listen to and argue with each others' positions, and also to explore the question of whether there was any connection between our different methodological approaches and our, equally different, policy/political stances.

The big methodological debates were between intensive and extensive research, the epistemological positions that underlay them and the different notions of rigour that each of them assumed. The relation to policy/politics was argued (by some) to derive from the fact that some approaches (extensive) took the capitalist nature of production as given while others (intensive) strove to conceptualize it explicitly. The question was whether the former approach led to 'policies' which sought to tinker with the system (taking it as given) while the latter tended to favour a 'politics' that addressed the nature of production more systematically. Maybe this reflected the times, before the neoliberal variant of capitalism had become so hegemonic. For if the methodological contest has, for now, been won by intensive research it is not clear that its subsequent development has retained the radical, system-questioning character.

Yet those questions of politics and method remain important, and it is good to welcome this new book. *Politics and Method* was a comparatively small venture. We were all from the United Kingdom, but we met face-to-face and tried to debate our differences without defensiveness or antagonism. The book reflected this – it was a handbook to be used. *Politics and Practice in Economic Geography* is bigger and rangier (though as its editors say even now still mainly Anglo-American), and its extent and variety is an indication of the major changes that have taken place in economic geography over the period. We are delighted to have the chance to welcome it, and appreciative of its reflections back over the period to *Politics and Method*. Being clear about method, still more actually *debating* method, continues to be important in all the ways the editors point to, from being more exacting about our own rigour, to forcing us to address the question of what is our political purpose within this field of 'economic geography'.

Doreen Massey, Department of Geography, Open University, UK.
Richard Meegan, European Institute for Urban Affairs, Liverpool John Moores University, Liverpool, UK.

Cover collage by Holly Peck, based on an original design by Andrew Robey for *Politics and Method*, 1986, Methuen.

PREFACE

Adam Tickell, Eric Sheppard, Trevor Barnes and Jamie Peck

Politics and Practice in Economic Geography is centrally concerned with research methods in the dynamic and heterodox field of economic geography, and the related social science disciplines with which it is increasingly closely embedded. But it is also a somewhat unusual book. Rather than explicating or advocating a particular set of research methods and techniques – the standard fare of research methods books – we attempt a bolder, and rather more unsettling, approach that lays bare the *practice* of using research methods to understand the social world. We take it as axiomatic that research methods matter – they are, after all, the means by which economic geographers (and others) encounter and explain the world; render research findings transparent and communicable; define their research objects and engage with research subjects; establish positionality; bring in political and moral views; connote the community of practitioners with whom they are associated; and make statements about the discipline they constitute.

Given this, it is remarkable how little attention methods have received in economic geography. A handful of insightful methodological interventions have episodically sustained the conversation, but too few economic geographical contributions are explicit about their methodological underpinnings or the ways that these influence their knowledge claims. These silences have come at a price: insufficient methodological transparency can get in the way of collective learning; the verification and triangulation of knowledge claims becomes more difficult; and communication and exchange with cognate fields, indeed across the field itself, are encumbered. Put baldly, the lack of explicit conversations about research methods has contributed to economic geographers talking past, rather than with, one another. It has probably also impeded the discipline's capacity to communicate effectively with cognate social-science fields.

We seek here to reinvigorate methods talk. Our aim in this book is to reflect on, and learn from, the interrelationships between research methods, politics and research practice in economic geography, thereby disinterring a range of long-buried methodological discussions. The inspiration for this volume came from a previous attempt to initiate just such a disciplinary conversation – Doreen Massey and Richard Meegan's *Politics and Method*

(1985b) – which put methods questions on the table for the first time, in the midst of searching debates around industrial restructuring. We have also drawn inspiration from more recent conversations among feminist geographers concerning the intimate connections between methodology, theory, politics, and practice. In this spirit, we asked contributors to reflect on methods in the context of the wider field of research practice – inviting them to take up such issues as ethics, the politics of writing, researcher positionality, commitments to relevance and transformative action, conceptualization, meaning interpretation, and research design.

We also asked each contributor to lay bare their personal engagements with methods in their research, inviting them to place themselves in the story and think about their own agency as researchers. We were seeking discussions that really opened up the dilemmas and challenges of practice. Too often, discussions of method suffer from an elision between an understanding of methods as a set of tools for practising science that can be separated from the research, and the reality that methodological choices generally have strong political, personal and serendipitous aspects. Our choices are almost always shaped by our training and socialization into the discipline (and our desire to position ourselves as extending or contesting prevailing trends); and our field sites, like our research questions, invariably reflect pre-existing interests and biographies.

Many of the decisions we make about methods we seem to make 'in private', or we bury them deep in research funding proposals, footnotes, or dissertation appendices. Only sporadically do we ventilate methodological concerns in our published output. For these reasons, *Politics and Practice in Economic Geography* cuts somewhat across the grain of disciplinary practice, and deliberately so. It is an invitation to the discipline to deepen its methodological conversations, to give voice to questions of research practice. Given the pluralized and fast-moving character of contemporary economic geography, these conversations are overdue. It is notable, but hardly surprising, that the essays that follow vary widely in tone, as well as in methodological inclinations, justifications offered for the choices made, and discussions of how politics and methodological practice are – inevitably – intertwined. We are a diverse field, and these conversations are really only just beginning.

This posed some challenges to organizing the volume: we encouraged our contributors to be reflexive and bold, to say things they might not normally say, and this is precisely what they did! As a result, the essays we received frequently exceeded the boxes originally conceived for them – in all manner of interesting and challenging ways. We have organized them around four overlapping themes, though most of the essays could appear in more than one place: (1) Essays that problematize the shifting relation between researcher and the subjects of research; (2) those that interrogate a number of the ways in which practices are bound up with politics; (3) contributions that examine the ongoing role of quantitative practices, and their relationship to qualitative economic geographies; and (4) those that take up the intrinsically geographical challenge of moving between different field sites, places of 'analysis', and spaces of engagement. These themes run throughout the volume, and we also pick them up in our own selective survey of methodological transformations in economic geography that introduces the book.

This collection demonstrates that there is not, nor should there be, methodological consensus in economic geography. Beyond demonstrating rude diversity for its own sake, our intention is that the different positions articulated here form the basis for a critical re-engagement with questions of research practice and politics in economic geography, and beyond. More methods talk, we hope, will enable new kinds of conversations and new practices. Ultimately, these should build on *and learn from*, rather than setting aside, their predecessors. This is not just about producing new economic geographies; it is about producing them in new ways.

ACKNOWLEDGEMENTS

The impetus for this volume came from discussions at the Madison and Bristol meetings of the Summer Institute in Economic Geography. Here, economic geographers gather to debate and discuss issues that they don't always find enough time to talk about, including methods. And it was here that the idea of a different kind of *Politics and Method* for a different kind of economic geography first arose. We gratefully acknowledge core funding for past meetings from the National Science Foundation, the Economic and Social Research Council, *Economic Geography*, the Universities of Wisconsin–Madison and Bristol, and the Worldwide Universities Network, as well as scholarship funding from *Antipode,* Blackwell, *Geoforum*, Sage, and the *Singapore Journal of Tropical Geography*. The moral and practical support of Kris Olds and David Angel in these endeavours has never been less than critical.

Proceeds from the book will help support future meetings of the Summer Institute.

All four editors would like to thank their current and former colleagues and students who, over the years, have challenged and unsettled their methodological suppositions. Adam Tickell thanks the University of Bristol for providing him with a remarkable amount of space to pursue his own interests. Eric Sheppard acknowledges the support provided by residency at the Center for Advanced Study in the Behavioral Sciences, Stanford, CA. Jamie Peck acknowledges the support of the John Simon Guggenheim Memorial Foundation, the University of Wisconsin-Madison Vilas Associates program, and, not least, all the staff and inmates at Meriter and UW hospitals. And Trevor Barnes is grateful to the National University of Singapore, where he held a Visiting Professorship during the formative period of the book's assemblage.

We would also like to thank the contributors to the book for their forbearance in the face of a protracted editorial process, and Doreen Massey, Richard Meegan and Robert Rojek for their support for the project.

Adam Tickell, Eric Sheppard, Jamie Peck and Trevor Barnes

METHODS MATTER: TRANSFORMATIONS IN ECONOMIC GEOGRAPHY

Trevor Barnes, Jamie Peck, Eric Sheppard and Adam Tickell

In 1983, Doreen Massey and Richard Meegan invited a small group of industrial geographers to a workshop at the Open University (OU) in Milton Keynes to explore the connections between theory, policy, and method in their work. The participants, a mix of established and up-and-coming British researchers, were called upon to talk about not only *what* they were doing, but *how* and *why* they were doing it, their theory and their methods. While a consensus solidified around the *what* of economic geography – the problems of manufacturing job loss, the stubbornness of regional divides, the apparently inexorable decline of the British economy – little agreement emerged about the *why* and *how*, prompting a lively, albeit inconclusive, debate. For all the sense of a shared project, there remained marked differences in research methods. In *Politics and Method*, the resulting book and inspiration for the present volume, the editors reflected: 'There were points, certainly, in the debate where an impasse was reached, but as someone said at lunchtime, voicing the thoughts of us all, "I wonder why we've never done this before?"' (Massey and Meegan 1985a: 4).

In many ways, our book is an attempt to resume that lunchtime conversation. But it is a difficult task, running against the disciplinary grain. *Doing it* rather than *talking about it* has been the dominant intellectual culture. The immediacy and pressing nature of issues confronting economic geographers seem to militate against methodological talk. Even Massey and Meegan, writing within the eye of the storm that was deindustrialization and restructuring, hesitated on the threshold of their own book: 'Time ... seems too short for self-conscious reflection on the nature of the relation between theoretical, methodological and policy perspectives' (Massey and Meegan 1985: 2).

But their conclusion was that methodological reflexivity was nevertheless indispensable. It should not be *either* politics *or* method – opting either for the study of industrial restructuring or 'self-conscious reflection' – but both: politics *and* method. Tackling urgent political issues like deindustrialization and restructuring required appropriate methods and serious talk. The earlier, brief flirtation of British economic geography with formal neoclassical theory and methods (regional science) was over, and it had since (re)turned largely to empirical analysis of industrial location, relying on orthodox data sources like censuses and surveys. But much of this work was theory-lite, or if there was theory it was loose and imprecise. Massey dismissed this 'locational factors' approach as a mindless 'search for a high R^2', incapable of coming to grips with the scouring politics of contemporary industrial change (Massey 1984: 15). The book she wrote in response, *Spatial Divisions of Labour* (1984) transformed the field. It triggered one of the sharpest paradigm shifts in contemporary economic geography.

The OU workshop was at the cusp of two transformative moments – within the discipline of economic geography, and within modern industrial history and politics. It was an instant when politics and method both mattered intensely, and were conjoined. This was not the moment to skirt around methods, or to mouth accepted political nostrums. Both required scrutiny in the searing light of contemporary change.

More than two decades later, economic geography has passed through a series of far-reaching cultural, institutional, and relational 'turns', during which time its objects, subjects, and means of study have been repeatedly overhauled. But despite these tumultuous changes, questions of method – the how and why of research – have been only fitfully (re)considered. With a few notable exceptions, methods went back underground. In this respect, the promise of *Politics and Method* was never realized. Of course, research *practices* themselves changed. Partly prompted by Massey's work, economic geographers came to embrace intensive case studies and interview techniques, 'finding the global in the local', and later they ventured into discourse analysis, participant observation, action research, ethnography, and much more. But, strangely, this produced little methodological talk, even though the methodological transformations were as significant as any theoretical ones.

The aim of *Politics and Practice in Economic Geography* is to bring to the surface methodological issues that for too long have been buried. We have no expectation of consensus, but believe that conversation itself is important, and especially critical given the current variegated and pluralized form of economic geography. While there are reasons to celebrate this vibrant heterodoxy, the far-flung and varied character of the field means that it is all the more important to sustain, *indeed deepen*, intra-disciplinary discussion. Under such circumstances, the risk of methodological relativism, fuelled by mutual misunderstanding, is a real one. More methodological talk can help here. Not that it will resolve all differences over disciplinary purpose and practice, but it provides a language of engagement. On those occasions when methods talk rises to the surface, as it did in the wake of Ann Markusen's (1999) 'fuzzy concepts' paper a few years ago, we are reminded how much remains unspoken in the field. While some saw the move towards intensive, case-study research as the early days of a 'qualitative revolution', others interpreted these same developments as a drift *away* from rigour and relevance (see Grabher and Hassink 2003). For Markusen (1999: 870), the slide into fuzzy conceptualization and slapdash theory development was accompanied by falling standards of evidence and an intellectual climate in which '[m]ethodology is little discussed'. Whether or not one shares Markusen's conclusions, she is surely right to call attention to the serious consequences of eschewing methodological talk. Methods matter.

Here, we set the scene for the collection by providing a (necessarily selective) review of some key methodological developments in the field of economic geography during the period since the publication of *Politics and Method*. We identify a number of cross-cutting methodological paths, drawing attention to some of the landmarks, intersections, and dead ends. The coverage, of course, cannot be comprehensive, and inevitably it reflects the limits of our own Anglo-American cultural and linguistic reach. Moreover, the field of economic geography never evolved in a sequential, linear fashion, through a series of identifiable methodological phases. There is no grand narrative with a beginning, middle and end on which we can draw. And in many ways, we would not want one. But it does make telling the story challenging.

Lacking the frame of a single linear narrative, we pick up two parallel, but occasionally intersecting themes. One starts with political economy, where initial concerns with industry and employment have broadened into bodies of work on business networks, technology, knowledge, and transnational commodity chains. The other that starts with feminism and post structuralism, with research on embodiment and work, extending out into explorations of social reproduction, the household, and social identity. Both strands are political, but differently political: in political economy, the political looms large, it is about class and economic development, whereas in feminism and post-structuralism it is characteristically 'small', about everyday possibilities for political action, including though the research method itself.

Methodological talk I: industry, place, networks

Economic geography's slow emergence from the cocoon of *industrial* geography coincided with its methodological awakening. By the mid-1980s, the taxonomic classification, 'industry', the basis of so much in economic geography, was under severe strain. As the introduction to the definitive collection of the time put it, 'the simple concept of a sector as a collection of establishments producing homogenous outputs is increasingly being called into question' (Storper and Scott 1986: 11). This was in part because changes in the 'real economy' were confounding industry classification schemes, together with the methodology of the 'bounded' sector study. Not only did forms of vertical disintegration characteristic of 'postfordism' muddy the singular category 'industry', but there was also growing awareness that service sectors warranted attention in their own right (see Storper and Walker 1989; Walker 1985).

If subcontracting, the rise of services, and more generally the emergence of flexible production called into question the reliance on *taxonomic* classifications of industry, other changes called into question sustained industrial production itself. Deindustrialization swept like a pandemic through the industrial heartlands of North America and Western Europe from the late 1970s. The subtitle to Bluestone and Harrison's (1982) influential book on American deindustrialization captured the moment: *Plant Closings, Community Abandonment, and the Dismantling of Basic Industry.* These conditions engendered a new sense of political as well as analytical urgency in the work of economic geographers, radicalizing the field. Some joined the front line, defending jobs and working-class communities, and formulating alternative strategies in collaboration with labour unions, community groups, and municipalities.

The result was an unprecedented yoking of politics and method. A new and rapidly diffusing set of research practices focused on issues like workplace change and work reorganization (the 'labour process'), international corporate strategy, and political alternatives (see Lovering 1989; Peck 2002). If workers and their communities were suffering, then the focus, as Bennett Harrison explained, should be on the boardroom 'perpetrators'. Harrison and others insisted that research must be 'explicitly value-driven and [should] consist of causal analysis combining strong logic and solid evidence to combat prevailing truisms' (Markusen 2001: 40-41). This ideologically-infused 'activist intellectualism' was a far cry from the reserved and ostensibly apolitical industrial geographies that preceded it. After this radical turn, data would no longer speak for themselves and managers would no longer be blithely characterized as optimizers. Research was now understood as a *political* process, and managers were almost certainly up to something.

To establish exactly what, economic geographers delved *inside* the production process to figure out the particularities and consequences of deindustrialization. More than mere data-reporting units, firms were analyzed as sites of class conflict and strategic decision-making. The challenges were to discover how companies really worked, how their markets were organized, what their competitors were doing, and the locational consequences. This prompted economic geographers to turn to interviews with senior managers, to comb company reports, to talk to union representatives, and to draw upon increasingly sophisticated surveys.

The break with past economic geographical practice was marked, especially in Britain, where a largely atheoretical tradition was suddenly and comprehensively challenged by the arrival of Marxism and critical realism. The latter was especially critical of the 1970s 'locational factors' approach. In this tradition, as Fothergill and Gudgin (1985: 104) advised, 'case studies should normally be avoided because it is difficult to know the extent to which the cases are typical, and the variability between cases often obscures trends that are clear when dealing with aggregate figures'. Realists, on the other hand dismissed such aggregations as 'chaotic conceptions' that lumped together causally unrelated and qualitatively variegated phenomena according to their co-location or descriptive characteristics (like company size or sector). Instead, realists favoured the close study of causally defined groups (such as a group of firms linked through contracting relations or experiencing similar forms of technical change), delving beneath statistical artefacts like 'net employment change' to identify the driving forces and mediating mechanisms involved in *specific instances* of restructuring. Explanatory penetration and political leverage, argued realists, stemmed from the careful weighing of the necessary and conjunctural conditions that shaped each 'case'. It would therefore be wrongheaded to imagine that a case might be more or less 'representative' or 'generalizable' (Morgan and Sayer 1988).

This was not, as Sayer (1985: 27) argued, the embrace of complexity for its own sake, or a turning away from abstract theory. Instead, it was a methodological counter to the widespread problem of 'apply[ing] theory at the concrete level without due regard for contingent mediations'. The intention was to sharpen methodological practice within a radically inflected 'new industrial geography'. In Sayer's view, however, even the new wave of work was cluttered with misleading stereotypes, empirical blindspots, and poorly formulated abstractions. It often amounted to the reduction of the concrete to the abstract, in effect fetishizing some forms of restructuring as indicators of a supposedly singular development path, while ignoring or marginalizing others. Sayer's realist injunction was not to question theory *per se*, but to utilize theory more carefully and reflexively.

The advice was to select case studies carefully, to cast the empirical net across a diverse range of circumstances (so as to *challenge* theory), and to engage in forms of *intensive* research marked by constant tacking back and forth between (provisional) theoretical explanations and cases, revising explanations along the way. This seductive formulation of the benefits of intensive research design was established by way of a pejoratively blunt, binary comparison with the 'extensive' methods that were 'far more common in economic geography' in the period prior to the realist ascendancy (Sayer and Morgan 1985: 150). Extensive research designs that sought to identify recurrent or 'typical' features of taxonomic groups, like firms in an industry or region, were damned with the distinctly faint praise of being

TABLE 0.1   **Intensive and extensive research compared**

	Intensive	Extensive
Research question	How does a process work in a particular case or a small number of cases? What *produces* a certain change? What did the agents actually *do*?	What are the regularities, general patterns, distinguishing features of a population? How widely are certain characteristics or processes distributed or represented?
Relations	Substantial relations of connection.	Formal relations of similarity.
Types of group studied	Causal groups.	Taxonomic groups.
Type of account produced	Causal explanation of the production of certain objects or events, though not necessarily a representative one.	Descriptive, 'representative' generalizations, lacking in explanatory penetration.
Typical methods	Study of individual agents in their causal contexts, interactive interviews, ethnography. Qualitative analysis.	Large-scale survey of population or representative sample, formal questionnaires, standardized interviews. Statistical analysis.
Are the results generalizable?	Actual concrete patterns and contingent relations are unlikely to be 'representative', 'average' or generalizable. *Necessary* relations discovered will exist wherever their relata are present, e.g. causal powers of objects are generalizable to other contexts as they are necessary features of these objects.	Although representative of a whole population, they are unlikely to be generalizable to other populations at different times and places. Problem of ecological fallacy in making inferences about individuals.
Disadvantages	Problem of representativeness.	Lack of explanatory power. Ecological fallacy in making inferences about individuals.

Source: Sayer and Morgan (1985)

useful at a descriptive level. The 'extensive method lacks explanatory pen-etration', Sayer and Morgan (1985: 152–3) disparagingly observed, 'not so much because it is a "broad–brush" method and insufficiently detailed, but because the relations it discovers are formal ones of similarity, dissimilarity, correlation, etc., rather than substantial causal relations of connection'.

The summary table of the two research designs (Table 0.1), which was faithfully reproduced in the methodology chapters of countless PhD disser-tations in subsequent years, might just as well have been labelled new wave/old hat rather than intensive/extensive. Advocates of orthodox, extensive methods within industrial geography apparently gave up without much of a fight. Intensive methods, with varying amounts of the supporting baggage of critical realism, became the new orthodoxy. This release of the methodological handbrake set economic geographers off on a search for theoretically informed case studies. Armed with new techniques like in-depth interviews, their mission was to divine causally meaningful, necessary relations and tendencies amidst the mass of contingent relations. No longer was there a singular empirical 'reality' whose inner secrets could be appre-hended by way of direct observation. The critical-realist world was a 'lay-ered' reality of structural/necessary relations, mediating mechanisms, and contingent outcomes.

While many, if not most, realist researchers drew their primary inspira-tion from Marxist theories, it was nonetheless an arm's length relationship. And when the criticisms of realism came, they were not from the positivist 'old guard', but from advocates of more hardened forms of Marxism, some of whom thought they could smell an empiricist rat jumping from the ship. David Harvey, for one, railed against what he portrayed as a take-it-or-leave-it approach to theory, posing as 'a convenient cover or as a transitional argument back to straight old fashioned and casual empiricism' (1987: 368). Sayer countered that while most of the protagonists in this debate were on the same side politically, this did not mean accepting Marxist arguments as a matter of 'faith'. Massey and Meegan, the Lancaster Regionalism Group, and other fellow travellers, he insisted, were elaborating theoretical expla-nations, not abandoning them. Sayer (1987: 398) feared that Harvey had begun to sound like 'an Antarctic explorer, not daring to wander more than a few yards from the tent of *Capital* for fear of being lost in an empiricist blizzard'.

Convinced that they were heading the right direction, the undeterred British realists soon ended up straying even farther from base camp. Fine-grained analyses of restructuring became the order of the day. A special pre-mium was placed upon what Sayer called the 'hard work' of concrete, empirical research, typically at the local scale. In the UK, the Changing Urban and Regional Systems (CURS) programme set out to elucidate the role of 'place' in the process of economic restructuring through the inten-sive study of seven British localities. Each was conducted by a separate research team, working within a shared methodological framework. If the localities programme set out to document some of the ways in which eco-nomic restructuring processes were 'embedded' in the matrix of local social relations, it did so as an exploratory and relatively 'open-ended' exercise (Cooke 1986). There was a unifying concern with the effects of gender relations (reflecting the growing influence of feminism within British eco-nomic geography), with local political cultures, and with a determined inquiry of grounded relations. 'The locality seemed to offer something concrete rather than abstract', Savage et al. (1987: 28) reflected, 'something specific rather than general, and a sphere where human agents mobilized and lived their daily lives'.

The localities initiative was immediately controversial. Neil Smith criticized this 'empirical turn' for merely

> documenting the minutiae of local change. ... The signs are already ominous. If indeed the unique is back on the agenda, then it is difficult to see how we can avoid fighting the crude Hartshornian battle between the ideographic and nomothetic, the unique and the general. ... Should this debate emerge again, as indeed it threatens to do, it will be tantamount to admitting that we have learned nothing since the 1950s (Smith 1987: 62, 66).

In retrospect, the methodological bark of the localities programme was much louder than its explanatory bite. Scott (2000: 28–9) subsequently characterized it as a 'rather straightforwardly descriptive and empirical research program ... an amalgam of inductive studies of regional economic trends and of commentary on the cut and thrust of local politics'. In his overview of economic geography's 'great half century', Scott gave the localities programme a place, but positioned it as an 'interlude'. David Harvey (1987) was less charitable.

While the localities initiative could claim some achievements – augmenting the standard operating procedures of industrial geography with more nuanced historical analyses (Beynon et al. 1994) and acute readings of gender relations (Bagguley et al. 1990) – much of the programme looked more like an explication of extant theory, enriching the catalogue of cases, than an effort to generate new explanatory insights. The basic methodology of industrial restructuring studies was expanded 'outward' to encompass a broader spectrum of local social relations, but this was achieved with the same methodological toolbox – semi-structured interviews and archival research, supplemented by secondary analysis of employment data. The locality studies lacked the ambition, depth, and salience of what in sociology were known as 'extended case studies' (see Burawoy 1991), and none rose to the level of 'critical case studies', selected for the potentially definitive challenges they pose for extant theory (see Mitchell 1983). Economic geographers were stepping outside the workplace, but the very richness of these studies of contextualized industrial restructuring came at the price of explanatory traction and conceptual sharpness. And as diverting as tales of manufacturing decline in the Birmingham suburbs may have been, they paled in comparison with the new wave of studies emanating from California. Interlude, indeed.

In contrast with the narratives of loss that pervaded the locality studies, the upstart California school was much more concerned with the dynamics of gung-ho growth in what were styled as *new* industrial spaces (Scott 1988b). Methodologically, this was back to the future. Again, the production process was seen, unambiguously, as the 'engine' of regional transformation (see Storper and Walker 1989). Allen Scott's early 1980s studies of proactively restructuring sectors like printed-circuit production, animated film-making, and women's garment manufacture drew on orthodox data sources like mail surveys and trade directories. Couched in an ambitious transactions-costs framework, Scott's sector studies served the modest methodological functions of 'corroboration', illustration, and empirical 'insight' (Scott 1983: 248). His prime motivation was the 'burning need to reopen macrotheoretical questions about the logic of capitalist society' (Scott 1988a: 183), a consequence of the systemic shift from Fordism to flexible accumulation. The economic crises of the 1970s had triggered

> a corresponding crisis of urban and regional theory. The old accounts of the forms of spatial development associated with the regime of Fordist accumulation and its cognate mode of Keynesian welfare-statist regulation were patently no longer very satisfactory as descriptions of underlying realities. ... One of the evident responses in human geography to this predicament has been a certain disillusionment with theoretical work in general and a radical return to empirical investigation, the multiplication of case studies, and an insistence on the significance of the local at the expense of the global and universal. ... It is my hope [to find] potentially fruitful avenues of reconciliation between theoretical and empirical work in human geography (Scott 1988a: 182–3).

Scott's forthright claims yielded a sharp response from John Lovering, an advocate of critical realism and a researcher on the CURS programme. Lovering (1990) took issue with what he saw as Scott's narrow, formal, and economistic theory of the firm – as a nexus of transaction costs. Scott underestimated the breadth of restructuring repertoire available to firms, focusing instead on 'sectors that fit [his] stereotype of postfordist flexible accumulation' (Lovering 1990: 167). Scott's (1988a: 181) confident identification of 'a common underlying system of structural dynamics' was, Lovering insisted, a byproduct of a parameterized, firm-centric analytical gaze, one that seriously underestimated (often divergent) local circumstances. Local outcomes were not determined by inexorable capitalist dynamics, but were socially and politically mediated, requiring concrete study.

Scott's response was no less robust, demarcating where the methodological battle lines lay. Lovering was fixated, Scott (1991: 133) countered, on 'indeterminate complexity'. The ultimate outcome of Lovering's 'methodological ponderings', Scott said, was a conception of 'empirical reality as just so much flotsam and jetsam, with theoretical ideas (so far as they play any role at all) reduced to inert atmospherics in the distant background'.

Lovering's wide-focus concern with the theorization of constitutive social relations, and with macro-institutional and local cultural contexts, was likewise dismissed as a forlorn pursuit of 'unstructured empirical investigations as virtues in themselves', bringing about a 'chilling effect on high-risk theoretical ventures'.

In the end, relatively few economic geographers heeded the call to high-wire theorizing; most felt more comfortable nearer the ground. But Scott's work on transactions costs nevertheless paved the way for the next generation of research, focused on a new object: the network. And soon, formalistic notions of industrial networks were giving way to more sociological treatments, turning on embeddedness and governance (Grabher 2006). The insight here was that relationships among elements within a network are governed by, and embedded within, a set of institutional norms, practices, and imperatives. Following Polanyi (1944: xlviii), embeddedness was the means to 'grasp [institutions] in their concrete aspect'. This concern with the nitty gritty of institutional forms, functions, and associations, meant that empirical skimpiness was absurd. To stint the concrete would be to undermine the very justification and object of study. Methodological challenges must be confronted head on.

Network approaches seemed uncannily suited to the question of explaining high-tech growth, a key concern for 'post-deindustrialization' economic geography. Animated by circuits of specialized knowledge, high-tech growth was evidently dependent upon embedded social networks of interpersonal relations and 'trust', and (re)produced through institutional governance structures, both formal and informal. Saxenian's *Regional Advantage* (1994) compared embedded governance structures and associated networks for two iconic high-tech regions: Silicon Valley and Route 128. These were indeed 'critical cases', capable of generating new theoretical insights, rather than merely illustrating extant theory claims. Breaking new ground for the discipline, the research strategy was overtly 'ethnographic in nature', with the empirical data 'accumulated over the course of nearly a decade living in and observing the two regional economies' (Saxenian 1994: 209). Sustaining the book's ambitious claims called for an up-close and personal engagement with sites and subjects. This was no quick and dirty study, nor could it be, given its comparative nature, and the variant of the network approach used.

Meric Gertler also derived conspicuous benefits from 'being there', from sustained and comparative case-study research. His work on advanced manufacturing systems explicitly begins with 'deep differences in culture' between Germany and North America (Gertler 2004: ix) – sufficient in some cases to bring work grinding to a halt. Like Saxenian, Gertler opts for a strategy of immersion.

Perhaps I am just a bit slow to figure things out. But I would argue that, from a methodological standpoint, there is considerable value in taking the time to immerse oneself in the different local contexts that ground business relationships in today's global economy. There is also real value in following the evolution of relationships and practices over time, as I have been able to do in this extended study. The second or third interviews I have been fortunate enough to conduct with many of the firms included in this study have often yielded the most useful insights. ... [There is no alternative to the] hard work of detailed, finely textured case studies: there is really no substitute for 'being there' (Gertler 2004: x).

Explanation here occurs through cases, richly documented in a fashion reminiscent of extended case studies. But in contrast to the work of both British restructuring researchers and the California school, the theoretical commitments are mostly implicit. The methodologically substantive contributions of Gertler and Saxenian seem much more likely to stand the test of time, however, than economic geographers' interventions in the related industrial districts debate. While rhetorically bold, these explorations of place-based networks were generally modest in terms of empirical substantiation. Not only did they remain within a tight productionist groove, they were also methodologically limited. Industrial districts could have been economic geography's finest hour – a construction that was constitutively geographical, and one that sparked interest across the social sciences. But in retrospect, the classic (or critical-case) studies of industrial districts of the 1980s and 1990s were carried out by economists, political scientists, and sociologists. Economic geographers missed an opening, and the sneaking suspicion is that methodological limitations were part of the problem.

Characteristically, economic geography moved on. Convinced that it was mistaken to sequester network relations to the local scale, economic geographers began to shift their attention to cross-scalar, *globalizing* networks. Again, much of this work was theoretically propelled, rather than empirically fuelled, though some of its enduring traces include work on Chinese business networks (Olds and Yeung 1999) and distinctive 'worlds of production' (Storper and Salais 1997). In the process, however, there was barely a methodological pause, and rare exceptions to this rule were as much concerned with analytical as methodological frameworks (see Dicken et al. 2001). The exacting methodological debates around the parallel project of 'global ethnography' (see Burawoy et al. 2000; Eliasoph and Lichterman 1999; Gille and Ó Riain 2002) represent a striking contrast.

Major impetus for economic-geographical work on networks came from actor-network theory (ANT). With roots in the sociology of science literature, ANT radically broadened the network optic. Now, anything could be an actor, from an INTEL micro-processor chip to a post-it note or a stressed-out day trader. Nothing is too small, intangible, or simple.

Indeed, the mundane and the everyday keeps the network functioning. For example, Leyshon and Thrift (1997) demonstrate the role of non-human agents, especially machines and texts, in their network account of London's financial centre.

The point of these arguments is that it takes hard work to sustain a network. At any moment, it may unravel, requiring the continual shoring up of old relationships, and the making of new ones. It follows that actor networks make their own geographies. They are not pegged on a pre-existing Euclidean frame; they generate their own topologies and geographical forms. And because of interaction between an agent and the network ('translation' in ANT language), new objects are continually created – known as hybrids or quasi objects. A good example is 'overseas' Chinese business networks, a hybrid object lying between traditional Chinese (*guanxi*) and Western business forms, but reducible to neither (Yeung 2004). As a *methodology*, ANT approaches are liberating in the sense that they privilege following networks, wherever they lead, while staying close to the ground (see Murdoch 1997). This is often seriously disruptive of extant explanations, and indeed this is partly the point: as Latour (2003) once gnomically pronounced, ANT is a 'negative methodology'. ANT's 'stories' are made in the telling; they do not follow a methodologically prescribed path. Methodological guidelines themselves might even be considered somewhat contrary to the project.

More methodologically prescriptive is the commodity-chain approach. At first blush, the commodity-chain concept seems clunkier than that of actor networks. A chain implies linearity, and the focus on the commodity seems unduly restrictive, but as economic-geographical extentions of the commodity chain approach increasingly allow for dense, multifarious, non-sequential, and dynamic relations, there is more fluidity. For example, Hughes (2000: 178) describes a commodity chain as a 'web of interdependence'. And since Marx it has been recognized that there is nothing restrictive about a commodity: 'a commodity appears, at first sight, a very trivial thing, and easily understood. Its analysis shows that it is, in reality, a very queer thing, abounding in metaphysical subtleties and theological niceties' (Marx 1976 [1867]: 163). Economic geography's best work on commodity-chain analysis tries to recoup those subtleties and niceties, revealing social, cultural, political, and not least geographical processes along the way.

Given the potentially enormous reach of commodity-chain and ANT analysis – comparison across national borders, much travel, use of translators, knowledge of disparate actors connected through the chain, understanding of an entire business, from resource extraction to retailing – the methodological challenges are demanding. But yet again comparatively little has been written about its method. A notable exception is Ian Cook's

(2001, 2004) work on the papaya commodity chain, beginning with plantations in Jamaica and ending as dessert in a North London flat.

Cook follows his commodity from the field to the plane, to the London supermarket, and to the fruit bowl. Although the narrative appears linear, it is not, as cross-cutting connections are drawn among and outside its constitutive stages: to colonialism, to the World Trade Organization (WTO), to Western middle-class consumption aesthetics, to the diseases and pests of exotic fruits, to the character of international air cargo transportation, and to developing-world labour control and surveillance. This is an unbounded, dense network of associations. And precisely because it isn't a simple chain, the commodity itself is no 'trivial thing'; it is composite, defetishized, decrypted, reflecting all manner of trace effects. Methodology is hardly trivial, either. Cook reveals problems in just getting to Jamaica, of keeping informants on side, of taping interviews, of translating and transcribing them. Even after he returns to England, the problems continue. There are rows in his department over his research, a shortage of money, and a disastrous presentation of his results, involving accusations of paternalism and jingoism.

Reflexively wearing one's methodological heart on one's sleeve like this can be disconcerting. It certainly represents a radical break from economic geography's conspiracy of methodological silence, marking the distance the discipline has travelled since the days of the aridly dispassionate 'sector study'. Yet the legacy of methodological reticence remains, particularly in comparison with those social science disciplines, like anthropology and sociology, that seem obsessively gripped by the need to reveal all. Surely, there lies something between exhibitionism and covering up?

Methodological talk II: praxis, feminism, post-structuralism

Political economy is one thread weaving through the last quarter-century of economic geography's methodological history. There is at least a second, sometimes frayed and broken, which starts roughly at the same moment, but which stresses different elements, draws on other intellectual traditions, emphasizes alternative political ends and means, and deploys distinct research methodologies. Especially at the beginning, it struggles to separate itself from political economy. But as it emerges, its theoretical sources become increasingly distinctive, primarily feminism and post-structuralism. Its foci are work, bodies, social reproduction, households, gender and ethnicity, and community. Interpreted through the lenses of discourse and power relations, analysis tends to be anchored in very particular geographical sites, both public and private. The nature of politics and political intervention is also different from political economy. Less couched at the scale of capitalism as a whole, it stresses active, local intervention, one means of which is the research process itself.

While there are these two separate thematic threads, they sometimes twist and knot, even in the lives and works of particular individuals. Doreen Massey, for example, made formative contributions to both. Indeed, she may not see any separation, but just a single thread. That said, for our purposes of trying to impose a narrative order on the disorderly methodological history of economic geography, it is useful to recognize the two traditions. Appropriately, then, we begin at more or less the same point …

The economic dislocations of the early 1980s not only helped to generate abstract, academic political-economic schemes like the restructuring approach. They also stimulated hands-on political action, focused on the immediate and the local – resisting plant closures, fomenting community action, serving on advisory committees and local councils. At least initially, praxis in economic geography seemed joined at the hip to political economy. It meant walking picket lines by day, reading Marx by night. In the United States, Bluestone and Harrison played a key role, influencing a generation of economic geographers (see Antipode 2001). Harrison's style of research was characterized as both action-oriented and collaborative, 'work[ing] closely with union and community-based organization leaders to craft research that would be useful to them', thereby establishing a research agenda 'not wholly shaped by the academy' (Christopherson 2001: 32). Indeed, Bluestone and Harrison's *The Deindustrialization of America* (1982) started life as a project commissioned by a labour–community coalition, both its 'analysis and its strongly worded policy recommendations [owing] much to what Ben learned from labor leaders, rank-and-file trade unionists, and community activists' (Markusen 2001: 40).

In Britain, from the mid-1970s, action research projects were initiated by the Labour Government's Community Development Programme (CDP). The CDPs spawned a series of community-based research projects, often triggered by events like factory closures or large-scale redundancies. It was in this context that '[e]conomic geography, and the questions of industrial location and labour markets in particular, took on a new political salience' (Lovering 1989: 201). Out of this ferment came, among other things, direct challenges to Thatcherism: the municipal socialist movement, and a closely intertwined intellectual project animated by the question, 'In what sense a regional problem?' (Massey 1979). There was also a significant, two-way movement – both of ideas and individuals – between the academic world and local government research units, community development organizations, and trade unions. One of the more notable outcomes of these efforts, alongside countless local jobs plans and municipal economic strategies, was the *London*

Industrial Strategy (GLC 1985), which 'drew heavily on the new industrial geography to propose fresh, practical ideas for the reindustrialization of inner London' (Storper 1987: 593).

Such work, both in the USA and the UK, was predicated on the belief that the larger economic system could be fixed. Through a series of reformist measures framed often at the local level, sufficient band aids could be applied to patch up the body economic. These initiatives, originally proposed by communities, organized labour, and progressive local governments, were by the late 1980s systematized by economic geographers, as well as by others, as full-blown schemas, plans and policy initiatives. They went under such rubrics as flexible specialization, postfordism, local competitiveness and 'cluster' strategies and, more recently, learning regions and creative cities.

J. K. Gibson-Graham – students of Ben Harrison, involved early on with community organization around plant closures – later reflected that the early 1980s was a period in which 'most leftists struggled ... without a clear vision of the future'. They fell into the belief either that local strategies were justified on the grounds that the transcendence of the 'system' was effectively beyond reach, or that 'the only way to create a non-capitalist alternative is to create a prosperous capitalism first' (Gibson-Graham 1996: 164). For Gibson-Graham, however, it was imperative to have that vision, accompanied by an appropriate theoretical and methodological armature. The forms of activism associated with political economy, and which later morphed into policies of postfordism, cluster strategies and the like, were not achieving what they might. This was because of the baggage of essentialism and totality that political economy carried, limiting what it could accomplish. In Gibson-Graham's view, political economy delineated 'economy' by a set of definitive characteristics, established in opposition to the social and the cultural, and envisaged as total, unified and pervasive. Consequently, where activism could take place, how intervention could be carried out, at which sites and scales, and by which actors and agencies were all narrowly determined by this essentialized conception. That conception needed to be overturned.

Gibson-Graham's project was to think about the economy, and thereby what it meant to fix it, in radically different ways. Their starting point was feminism and post-structural theory. Both offered a more expansive and porous conception of the economy, pointed to topics consistently underplayed or ignored, emphasized a richer, more varied palette of research techniques and questions, and indicated alternative forms of praxis.

Feminism first entered economic geography around the moment of deindustrialization, in the form of empirical studies of the spatial constraints women faced in the employment market. By the mid-1980s, those studies were joined by theoretical discussions around domestic labour and patriarchy taken from Marxism (socialist feminism). By the late 1980s, there was yet another break, stemming from changes within feminist geography and the introduction of post-structural theory (Bowlby et al. 1989). That theory was multifaceted and dynamic, but a central claim turned on the construction and power of discourse in shaping, permeating, and directing non-essentialized differences of all kinds: gender, race, subject identity, bodies, and, as we shall see, even the economy. Moreover, this was theory designed as Foucault said 'to groan and protest' within concrete studies. It was no armchair theory, but gained legitimacy at ground level from the deployment of situated research methods. Finally, it was theory with a political point, to change the world. Unless strategies for political change were part of its furniture, it was irrelevant theory.

This combination of sensibilities began to transform economic geography from the early 1990s. Doreen Massey's (1995; Massey et al. 1992) work on the high-technology complex in Cambridge (UK) was among the first to demonstrate the importance for economic geography of focusing on socially inscribed bodies. With Nick Henry, not only did she talk to economic geography's usual subjects – company directors, personnel managers, key workers – she also 'interview[ed] the partners of the scientist/engineers in the cases where they were cohabiting' (Henry and Massey 1995: 50). What they found was workaholism run rampant, which affected more than just the workaholics themselves. It impacted on those around them, such as administrative assistants, secretaries and partners. Unsurprisingly, there was a clear division in the way in which the bodies involved were socially marked: more than 90 per cent of the high-tech workers were white men, mostly under 40, and graduates, while most of the assistants, secretaries and partners were women.

Massey's point was that while scientists and engineers portrayed themselves as disembodied rational minds, ferreting out objective truth in the name of science, the organization of their corporeal lives told another story. The inordinate hours that men worked as scientists was made possible by work of a different sort, done by their predominantly female partners and support staff (Massey 1995). Male abstract rationality was achieved by female labour.

In many ways, Massey's study lay between political economy, and feminism and post-structuralism. Still centrally concerned with class and the economy, making use of a traditional large-scale industry survey as well as interviews, it also pointed to a different trajectory, focused on gender and work, the construction of masculinist discourses, and intertwined relations of power. In addition to asking conventional questions about production, it also problematized reproduction, asking 'who irons the shirts?'

Yet more ambitious forms of methodological fusion were to follow. Hanson and Pratt's *Gender, Work and Space* (1995) married established concerns of feminist labour-market and transportation studies with post-structural theories. The book presented a complex layering and interlocking of different methodologies. There are certainly tables of numbers, of travel times, wage rates, gender employment ratios, and even the occasional chi-square statistic, but there are also rich, complex, contextual stories about identity, mobility, constraint, and local situatedness, gathered through focus groups, extended interviews and ethnography. The strength of the book lies in the braiding of these different methods into a coherent narrative about the nature and effects not only of gender differentiation, but differentiation within gender, along the fault-lines of class and race. Methods here are not mere instruments, external tools to achieve a particular end, but are integral to the end itself.

And the feminist injunction here, beyond recognizing the often concealed nature of 'women's work', is to see how economic relations are themselves co-constituted through gender relations. In a radically different context, this was also a defining theme of McDowell's (1997) examination of the performance of merchant banking work in the City of London. Here, the body itself constitutively entered into the economic product, in this case, financial services:

> The theoretical focus on the body, sexualised performances and strategies of surveillance parallels material changes in the nature of work in service sector occupations. One of the key features of service sector work, compared with manufacturing jobs, is that the labour power and embodied performance of workers are part of the product in a way that was not the case in the production of manufactured goods ... In service interactions, the body of the worker demand[s] an embodied and visible performance (McDowell 1997: 32)

Conventional accounts of workers in the City of London tend to privilege their (disembodied) mental capacities – the ability to make split-second calculations about buying and selling. They appear, like Massey's high-tech scientists, as minds without bodies. McDowell's argument, however, is that bodies are integrally involved in the product: workers must talk the talk, but also walk the walk. They must perform appropriately, following the script, looking the part. Bodily acts constitute the reality of the 'good' that service sectors sell. Economy incarnate.

Researching embodied practices is an embodied practice itself. McDowell's first sentence is: 'This book is the result of field work in the City of London' (1997: v). Interspersed with her 'results' are accounts of what it was like to be in that field, how she operated, how her own performance as a socially inscribed body affected her results, what types of methodological strategies she deployed and their justification, and why it mattered politically that the focus was gendered bodies and feminist methods. Like Hanson and Pratt's work, but unlike many economic geographies that preceded it, McDowell's methodology is never separate, implicit or invisible.

The point, in fact, had emerged earlier in a debate between McDowell and Erica Schoenberger about corporate interviews – a rare outbreak of reflexive methodological debate in economic geography. Schoenberger's (1991) defence of the technique of the corporate interview had been primarily aimed at a methodologically orthodox audience sceptical of such 'subjective' qualitative approaches. McDowell's criticism, however, came from a quite different direction. It concerned Schoenberger's lack of recognition of gendered bodies, hers and those of her interviewees.

> I wondered about her reason for not addressing one of the most significant factors in her own research – the fact that she is a *woman* and that almost certainly, the majority of her respondents were men. Earlier in her paper she suggested that in corporate interviews the investigator and the respondents were likely 'to share the same language and social background.' So why did she omit to consider the one single important difference between them – that of their sex? Was she so certain that gender relations were irrelevant in these circumstances? (McDowell 1992b: 214)

In her reply, Schoenberger (1992: 217) said: 'I agree with McDowell that gender makes a difference. The reason I did not talk about it was that I am not sure precisely what difference it makes, and I am not sure how I would know.' Her primary purpose was to expose cultural constructions of corporate culture, in contrast to the rationalist logic employed, not least, by managers themselves (Schoenberger 1997). As she would later explain, 'the corporation ... is both a site of capital accumulation and a stage for the playing out of powerful psychological and emotional processes. This suggests that we're not going to understand what corporations do in the world and why without analyzing this turbulent mixture of passion, power and rationality' (2001: 296). But both actors and analysts can have blindspots, and the purpose of McDowell's feminist intervention was to call attention to the constitutive role of gender relations in sites where they are so often frozen.

Gibson–Graham's (1996, 2006) work is likewise productively disruptive, but in explicitly political ways. The action–research method sought no less than the transformation of the subjects of study, and their circumstances. Gibson–Graham suggested that, rather than wait around for once-and-for-all, systemic change, progressive transformation could be achieved piecemeal, bit-by-bit. Furthermore, and this was the lesson of feminist methods, praxis could be achieved through the very research process itself.

Gibson-Graham's site was four coal-mining towns in Queensland, Australia. The subjects were the female partners ('mining town women') of the male miners working there. Characterizing prevailing household class arrangements as 'feudal', Gibson-Graham argued that the bodies of women had been given over to patriarchal domestic service, 'hewers of cake, drawers of tea'. Because of declining markets and falling prices in the late 1980s, coalmining companies pushed to 'remove restrictive work practices' (Gibson-Graham 1996: 225). Instituting the 'seven-day roster', companies made miners work more weekends, significantly increasing the workloads of female partners. This was the moment at which Gibson-Graham began to research. They did not want to carry out standard social science, interviewing the women, treating their stories as simply data for analysis, nor were they interested in another orthodox political economy, just another illustration of capitalist crisis and oppression. Rather, in the spirit of action research, they wanted to use the very process of interviewing mining-town women as a political opportunity for progressive intervention. Hiring mining-town women, first to participate in workshops and later to carry out interviews with other women in the community, the research project was carried out 'in workshops and over the kitchen tables where one-on-one interviews were conducted [a strategy that] created and cultivated spaces in which a feminist politics (the transformation of gendered power relations) was performed' (Gibson-Graham 1994: 215). In turn, this influenced male partners, ultimately making a difference in the mine itself.

This is a striking illustration of forms of political intervention in the economy made, as it were, below the waterline. The 'formal' economy, constituted through wage labour in capitalist enterprises, is but the tip of a much larger iceberg, Gibson-Graham insisted. A plethora of alternative economies exist in the mostly submerged 90 per cent (see Figure 0.1). One of the tasks of progressive, action-oriented research is to enlarge the field of the visible, not least with the assistance of research subjects themselves. The purpose is as much political as analytical: when they believe there is only one essential economy, uniform, pervasive and impregnable, people give up before a hopeless task. But by pointing out the possibilities of diversity, that structures are not universally given, change becomes possible. Giving a lie to the essential economy therefore exposes new futures.

Geraldine Pratt's *Working Feminism* (2004) is also about trying to realize new futures through the research process. In this case it is for Filipina domestic workers in Vancouver, who, she contends, are abused both by their employers and the Canadian state. As a feminist, Pratt insists that it is necessary to make lines of connection with women who are otherwise very different from her; as a theoretician, she thinks it is necessary to interrogate abstract claims against the stubborn materialities of a prolonged case study;

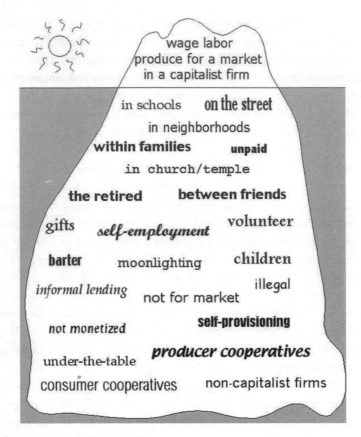

FIGURE 0.1   The economic iceberg

Source: J.K. Gibson-Graham (n.d.)

and as a methodologist, she is keen to explore how far one can take a single case study before repetition and banality take over.

At the heart of this extended case is collaboration with the Philippine Women Center in Vancouver. She began the co-operative relation for political and methodological reasons. Politically, she wanted to be part of an existing organization in which she could help facilitate progressive change for Filipina women predominantly employed as domestic labourers and caregivers. Methodologically, she recognized that

> sustained collaboration ... forced a responsibility to theorize in concrete ways. Working with any community group has this effect. ... Working with Filipino activists' groups is to be continuously reminded of the limits to generalization, and the value and necessity of theorizing within the everyday rather than viewing the empirical merely as an illustration of a vehicle for abstract theorizing. (Pratt 2004: 8)

Figure 0.1   Drawing by Ken Byrne; reprinted with permission from the authors.

So, like Gibson-Graham, Pratt uses both 'extended' research methods and prolonged forms of engagement to realize political ends. But these ends entail her 'continuously' having to rethink both methods and theory.

Both essentialism and masculinism run through the entire history of economic geography, shaping its content, delineating what counts as acceptable knowledge. First feminism and later post-structuralism substantially transformed the 'subject matter' of the discipline. Not only did the topical concerns of economic geographers shift, their conceptions of subjects, and relations with those subjects, also changed. Feminists, in particular, have been much more open to making methodological talk, showing what the discipline might be if, rather than skirting around its methods, it opts to expose, celebrate and strategically deploy them. Such a task is ever more urgent as the discipline multiplies and proliferates, respecting no boundaries, either substantively or methodologically.

Conclusion: talking methods

Feminist economic geography is the exception that proves the rule: for the most part, the discipline appears to operate a 'don't ask, don't tell' policy with respect to methods. We all use them; we just don't talk much about them. Sustained methodological reflection is exceptionally rare in published economic geographical research. Tracking down even cursory information on methods involves reading between the lines, even in methodologically creative texts. Economic geography is methodologically opaque. This alone, surely, is reason enough for economic geographers to begin talking about methods.

In the period covered by our review, economic geography was a fast and fecund discipline: theoretical frameworks and topical concerns turned over quickly, and research orientations and practices also were repeatedly transformed. The analysis of censuses and surveys, once the methodological norm in the field, is now a minority practice; formal modelling and sophisticated quantitative analyses are almost lost arts (see Hamnett 2003). Meanwhile, an array of new theoretical influences (feminism, regulationism, evolutionary economics, economic sociology and economic anthropology, non-representational theory etc.) and substantive concerns (networks, cultural embeddedness, embodiment, governance, social reproduction etc.) have led economic geographers into new methodological territories. Case-study research, once the exception, is now standard. And a wide range of qualitative and intensive research methodologies – in-depth interviews, ethnography, discourse analysis, oral histories, focus groups, participant observation, performance, activist engagement – all have a place.

But many of these approaches would have been considered avant-garde in economic geography as recently as the mid-1980s.

This remarkable period of *understated* methodological proliferation has been a creative and productive one for economic geography as a field. A number of recent 'state of the art' collections attest to the discipline's rude health (Clark et al. 2000; Lee and Wills 1997; Sheppard and Barnes 2000). So what's to worry about? Notwithstanding the progress made in the field in recent years, there has been a growing undercurrent of anxiety. Is economic geography losing its focus, its role and relevance? Is it asking the right questions? Is it connecting in the right way to policy-makers and other audiences? What is its (appropriate) position within the academic division of labour? What is its distinctive message?

The general paucity of methods talk in the discipline, facilitated by conventions largely of its own making, has done nothing to alleviate these concerns. In published works, the analytical rationale for case-study selection rarely receives more than cursory attention. By the standards of extended-case methodologies, the extension of cases through time (entailing protracted temporary commitments to field sites), over space (via multi-locale case studies), and across scale (connecting 'global' to 'local' processes) is only sporadically achieved. Sustained theoretical development by way of case-study research is hampered by an apparent disinclination across the field to invest in corroboration, triangulation, and interrogation across comparative sites. The degree of codification and 'sharing' of research practices, which would entail high degrees of transparency and reflexivity, lags the uptake of qualitative methods. Sampling strategies and procedures are often glossed over, particularly in low-n studies. It is rare to see autocritical methodological reflection in economic geography papers. With a few notable exceptions, issues of researcher positionality remain unacknowledged and unexamined.

Not only is much methods talk reduced to footnotes, the field of economic geography may also be somewhat vulnerable to the charge that its aggregate level of methodological experimentation and creativity is not high. Interviews are the method of choice across large parts of the discipline. This is an area where there is productive, if sporadic, engagement with issues of positionality, ethics, and technique (see Clark 1998; *Geoforum* 1999), but venturing beyond interviews is not as commonplace as it might be. Ethnography and participant-observation techniques occasionally are used, but genuine ethnographic depth remains illusive (Herbert 2000; Lees 2003).

This does not mean that economic geography is bereft of good methodological practice. Rather, it is a comment on the nature of our disciplinary conversation. Across the field as a whole, a low premium is placed on methodological codification, reflection, transparency, and reflexivity. The consequences include: a narrowing of the bandwidth of our disciplinary conversation; a slowing of the pace at which the 'tricks of the trade' are learnt by entrants to the field, who often must remake old mistakes and reinvent wheels as a result; and a reduced capacity to engage with other disciplines, which are less likely to extend the kind of methodological trust that we so freely share among ourselves.

It follows that, if economic geography is able to overcome its methodological reticence, there will be paybacks in a number of areas. The first and perhaps most obvious is an enriched and 'thicker' disciplinary conversation, in which research practices and processes are taken seriously. As the following contributions demonstrate, encouraging economic geographers to talk (again) about method is not self-indulgent soul-searching, it opens up *and sharpens* questions of politics, purpose, and priorities. Second, talking more about methods has the potential to increase the 'social productivity' of the discipline. Methodological transparency facilitates learning from others. Both good methodological practice ('minimum standards') and methodological creativity (pushing the envelope of research practice) are enabled by more methods talk. Third, the scrupulous articulation and defence of method both keeps us honest and aids rigorous conceptualization. Method provides a crucial 'hinge' between empirical evidence and theoretical claims. And fourth, methods talk facilitates more effective and deeper forms of communication, improving cross-disciplinary engagement, and reducing the risk of Balkanization.

A more methodologically reflexive economic geography will also be a *different* economic geography. There are certainly upsides to economic geography's disciplinary velocity, including its adventurousness, novelty, energy, and creativity. But there is also a downside to a 'don't look back' ethos. There is little patience for corroboration and confirmation; there is a preoccupation with the moment of disclosure, rather than the slower processes of substantiation and extension; and arguments are rarely interrogated in depth before the field moves on. Martin and Sunley's (2001: 153) commentary on the condition of economic geography, for example, reached the sobering conclusion that 'the majority of new concepts now go uncontested'.

In contrast, a more methodologically deliberative and reflexive economic geography would be a more *methodical* economic geography, focusing attention not just on staking claims but on *sustaining* them. Our aspiration is an economic geography that is as effective at holding its ground as it is in breaking new ground. And the ground should also be methodologically pluralized and *contested*, in contrast to the present situation, in which different islands of practice in the discipline tend to be 'mono-methodological'. Methods talk is not exclusionary, an official set of prescribed topics of dialogue or, even worse, a gagging order prohibiting certain utterances. We are not trying to shut anyone up; quite the reverse. Methods talk liberates and deepens the conversation, allowing us more effectively to connect politics and practice.

Section 1

Position and Method: Producing Economic Geographies

1 POLITICS AND PRACTICE: BECOMING A GEOGRAPHER

Erica Schoenberger

As one moves deeper into a subject, one sees how much the power of the subjective operates, even in the sciences; and one does not advance until one begins to know one's own self and character. (Goethe)

In this chapter, I try to work through aspects of my self and character in order to better understand the connections among theory-building, research questions, method and politics – at least in my own work. (Better late than never, you might think.) My approach will be to apply to myself the method I have used in analyzing corporate executives. This means trying to work out questions of identity, history, and social position and how these are involved in producing both ontological and epistemological commitments and a sense of the rightness and wrongness of things.

The difference between corporate executives and me, of course, is that their actions in the world are consequential and mine aren't particularly. So one might wonder what the value of this exercise is. I offer it here as a kind of heuristic. I think we should follow Goethe's advice, and am just barely willing to try to do it in public, if only to illustrate the astonishing misperceptions and horrible delusions that may complicate the task.

This kind of analysis, being retrospective, is unavoidably neater and more logical than the lived experience, which felt more like a long series of bizarre accidents. It had never, for example, even entered my head to become an academic until years after I was one. So much for clarity.

Identity, commitments and research project

I want to propose here that our specific historical and geographical origins critically shape our identities and worldviews and these, in turn, underlie the shape and orientation of our research. For me, the identity at issue is political identity – not merely being a leftist but being a *leftist* – this is who I am. But I am a leftist who came of age in the American anti-war movement of the 1970s.

This means, first of all, that I grew up in a place with no tradition of viable labour party or socialist party politics. For people who come out of the labour movement or socialist politics, I imagine that labour, work and class appear to be more or less 'natural' objects of research. For people coming directly out of the civil rights and/or feminist movements, race, gender, community and place may be more immediately salient. The politics that I grew up in was involved in the first instance with political and military power and oppression and connected this up with economic injustice and social class rather than the other way round. For someone like me, the neo-imperialist state, the defence industry and multinational corporations are what appeared most urgently to require attention.

A little bit more in the background, this historical context also produced persistent questions about the misuse of power linked with the misuse of knowledge. That is to say, knowledge about what ought to be done or what is feasible and effective is available and accessible, yet is contravened because those in power are pursuing contradictory goals. So a large and powerful entity (in this case the USA) ends up acting against its own best interests while causing a tremendous amount of damage to others. With the publication of the *Pentagon Papers*, it became abundantly clear that the government had consistently lied to its citizens about nearly everything connected with the origins and prosecution of the war in Vietnam. We also know that the government pursued strategies in that war that it knew in advance would be ineffective because it had to do *something* and couldn't think of anything else to do that would be compatible with its own idea of itself (McNamara and VanDeMark 1995; Sheehan 1988). The costs hardly need enumerating. Vietnam would stand as a monument to a certain kind of human and social folly, were we not actually engaged in repeating the entire process – complete with lies and the misuse of knowledge – in Iraq. Apparently, we cannot yet erect the memorials to this particular pathology.

I want to make two observations about this problem of power and knowledge. The first is that it took a while for it to enter explicitly into my research. For a good ten years I was wholly absorbed in the problem of how and why firms sought to be competitive and to maximize profits through locating in high-cost, highly unionized, and highly regulated places. I wasn't really attuned to the question of how and why they could also make staggering errors of judgement with hideous consequences for many people and places. This lack of attention was perhaps conditioned by my training in both conventional economics and political economy. In the former, individual capitalist rationality leads to rational and good social outcomes. In the latter, individual capitalist rationality leads to irrational and negative social outcomes. Neither theory is looking for irrationality at the firm level. And indeed, the problem is not exactly that the firm *per se* is irrational, but that the people running the firm have conflicting rationalities on behalf of themselves as social agents and on behalf of the firm as a social agent. I can see this now as analogous to the politically self-serving, identity-preserving yet nationally harmful behaviour of the Vietnam (and Iraq) war strategists. And perhaps this political conditioning allowed me to finally recognize it when I saw it repeatedly in corporations. But it took a while.

The second observation is that use of the phrase 'power and knowledge' seems to cry out for a Foucauldian analysis which I have never engaged in. This is for several reasons, based on a possibly defective reading of Foucault. One is that, for my money, power and society are rather reified concepts in Foucault. They have their own interests and objectives. These aren't, for me, clearly connected to particular group or class interests. In my ontological universe, it is necessary to understand *whose* power is at issue and why – given whose it is – it wants what it wants. I want the mediation of class and social position, which I find much more clearly and productively articulated in Bourdieu. Further, the reified power of Foucault seeks knowledge in order to discipline people 'in general' and in so doing stabilize the social order. I want also to understand something about how social position and power relate to knowledge of self and the ability to productively use knowledge in general. Or, put another way, I want to understand how social position and power inflect the transformation of information into particular knowledge and how exactly this knowledge will be used.

Back to me. For various reasons connected with my personal and family history, it was both easy and comfortable to choose a corporate and defence focus rather than, say, researching the state directly. My father achieved middle-class status via the military and worked in the defence industry all his life, specializing in international marketing for a number of large, lethal firms. So I am a child of the military–industrial complex and the utterly white-bread suburbs. I suffer a lot from red diaper envy. I also apparently organized my research in part as a way of explaining my father to myself. This I realized so recently it would make a cat laugh.

Beyond that, I worked my way through college and while doing political work for some years afterwards as a secretary, bookkeeper and production analyst, which is how I learned to talk to businessmen (and – key to securing interviews – their secretaries). I also learned something of how they think and what they really spend time thinking about. It turns out if you're a really good secretary (and I was), they don't treat you like a servant. They treat you as an extremely ill-paid junior partner. And along the way I learned how to construct and read financial statements.

I'm not arguing for a perfect isomorphism between historical–geographical moment and research trajectory. But I think it may help explain why particular issues emerge as the big problems for each of us. We are thinking about them, confronted by them, stumped by them (how do they get *away* with it?!), angered or otherwise impassioned about them all the time. It is what we *need* to do.

Project and practice: the question and how to research it

When I was a graduate student, discussions about deindustrialization and the New International Division of Labour were very much in the air (Bluestone and Harrison 1982; Froebel et al. 1980; Harvey 1982; Massey 1984). It seemed evident that manufacturing investment was destined to desert the USA and other advanced industrial economies. Indeed, on this subject, there was a curious and troubling convergence between the leftists and mainstream economists.

As I was reading this literature, though, I was also obsessively reading the business press, a habit I acquired working for different left-wing research groups. And in the business press, what I was seeing was a positive flood of announcements of foreign manufacturing investments in the USA. My first reaction was a kind of indignant exasperation: 'What idiots! Don't they know any better?' Then I thought it was probably better not to assume that they didn't know what they were doing. So the question to research seemed plain: what on earth are they up to?

It is childishly easy to show statistically that international manufacturing investment is highly correlated with high wages, unionization and regulation. But the statistical evidence could not produce *explanations* of what I was seeing. The only alternative seemed to be to go talk with the people who were making these decisions. With my background, this seemed both obvious and relatively easy.

We can't, of course, just ask them why and write down their answers. We're not reporters. Neither are we ethnographers, but that model is closer to the point. We don't want to list the practices the natives engage in. We want to understand them in the context of an entire way of life. So you talk with them about the whole way of life, and it is in that story that you find the explanation of the practice. You talk about the history of the firm, the history of the person, etc., and when you've got that in hand, then you ask for their explanations of the specific practices. These explanations are not the answers to your research questions – they are *data* that need to be analyzed and interrogated.

For a long while, I resisted any impulse to analyze the *people* who were providing this information. I'm very comfortable with large-scale political economy and the 'logic of capital'; the structure is my natural home. I knew I needed to talk with these guys, and to know enough about them personally to interpret the information they provided, but I didn't feel that I needed to *think* about them particularly. But what I had learned in investigating the structure (e.g., how competition works to produce particular social and geographical outcomes) is that firms often make hideous mistakes even though they apparently inhabit the high ground of capitalist rationality. So it wasn't quite enough to get at what the strategy was and the plausible structural explanation behind it. In effect, the failure of strategy forced me to look more closely at the strategists and how they and their strategies are produced.

This meant taking on the capitalists as individuals and as a class simultaneously, which was confusing and hard to work out. I turned to culture and identity as a way of managing this inquiry – as a way of thinking through it without going insane. They functioned for me as meso-level categories and conceptual anchors that connected the big structural dynamics (competition, production, investment, accumulation) with how a particular class of agents operated within those dynamics in a particular conjuncture.

So the first version of the work centred on the question 'why do firms work this way?' and answered it by saying 'their behaviour is driven by how competition works in these industries; the competitive strategy forces them to behave in ways that seem to be at odds with purely production-related considerations such as labour costs'. The second version of the work is about the same thing – what firms do and why – but in a different register. It tries to explain where competitive strategy comes from and acknowledges that it does not derive solely from analyzing how markets work in particular industries. What I gained through this shift, I think, is a richer notion of competition among firms, of intra-class competition and the problems and opportunities this creates for firms and workers alike.

But being fixated on the firm as the unit of analysis, competition as the central problem and strategy and culture as the source of movement in the system, also removes many important dynamics from my field of vision. I don't, for example, 'see' gender, labour strategies, or places. These are no small lacunae, and I would feel worse about them were it not for the fact that other people see them extremely well. The question, I suppose, is whether my way of seeing can be brought sufficiently into alignment with theirs that we can achieve real depth of field.

Politics and objectivity

Can a confessed leftist be rigorous, honest and fair as a researcher? I believe so. The politics help you identify the important questions, but they can't tell you what the answers are. In my experience, leftist academics are, in general, much more attuned to how their politics shape the questions they are interested in than their mainstream colleagues who cling to the notion that their research is 'value neutral'. So on the whole, I think we're ahead of the game.

Here one might ask: can I seriously believe that politics, questions, theory and methods – hence answers – are not deeply connected? I have to acknowledge that the answers I'm likely to arrive at in my research are not somehow quarantined from the rest of me. My politics inform my theoretical approach, out of which the method really grows. So the kind of results I am able to get are unavoidably shaped by this research *gestalt*. Simple honesty can keep me from fixing the results to accord with my preferences or political commitments. But with the best will in the world, I cannot erase myself from the process. Nor would I want to. However, what I do want is to make sense out of the world and in order to do that, one is bound to be as critical and self-critical as one has the capacity to be. This means, one hopes, wishful thinking is kept to a minimum. Or perhaps I'm entirely wrong about this and should just say, to paraphrase a famous US Secretary of Defense, you work with the self you have, not the self you wish you had.

Do the politics enable me to ask questions that wouldn't otherwise emerge?[1] Certainly a commitment to a political-economic ontology and epistemology, which for me is essential to being a leftist, opens up a characteristic set of questions that are not available or particularly interesting to mainstream economists. Rather than an obsession with how scarce resources are allocated by anonymous individual decision-makers, I am obsessed with how resources are produced and distributed in society, who controls them, what they do with them, why they do that, and how it affects everyone else. In a nutshell, I want to know about the creation, distribution and use of wealth and social power and how this is connected with how values concerning, for example, social justice and the environment are constructed and operationalized. There are more ways of getting at those questions than being exactly the kind of leftist/political economist that I am. Within this context, however, they are not only available but unavoidable.

Having said that, a political economic approach generally treats social power as a structural artefact, at least in reference to the ruling classes. It is more interested in the lived experience of power from the point of view of workers, and, plausibly, communities, racial and ethnic groups and women: those who are generally on the weaker end of power-laden relationships and who have to work at countering, resisting, deflecting or otherwise preserving themselves from the powerful. In my own work, however, I was focused on what the powerful were doing and why, and, because this wasn't so easy to work out, it led eventually to an interest in how the powerful thought about their world and about themselves – in short, how their actual power was linked to their identities, their interpretation of information, and their strategies. This kind of question doesn't, I think, automatically arise in a classic political-economic analysis, but it does plausibly arise in the context of the kind of political history I've been describing. The *particular* left politics of the USA in the 1970s were in fact quite involved with the question of who these people ('the best and the brightest') were and what they thought they were doing. The policies of the war in Vietnam were located in a social type – Ivy League whiz kids and technocrats – and closely identified with particular people – McNamara, Bundy, Colby, Kissinger, etc. It seems likely that this helped me find the path from corporate strategy to culture, via questions of power and identity.

Interlude: the work of mentors

There were plenty of reasons to doubt my career potential in academic life, quite apart from the fact that it took me so long to notice that I was having an academic career. Being a leftist, a qualitative researcher and a woman are not the first three indicators of success that leap to mind. In any case, most of us aren't so incredibly brilliant that the system just bows down before us. We're all peculiar in some ways, and we all need help.

This is what makes the work of mentors so important. What, then, do mentors do? They give advice, which is great. Their critical and unsung role, however, is to ensure that the system works fairly and well in your particular case. A lot of the work that the system does in the course of your career happens out of your reach – literally behind closed doors and involving people whose identities are secret. The system may be fair in principle, but it is not working on your behalf. It works for the institution. Mentors are your advocates and watchdogs in the work that the system is doing. I don't mean they fix it so you can't fail. I mean that they watch out for you. They make sure that obvious errors aren't made, that there are answers to questions, that inappropriate criteria or inappropriate people aren't used to judge you, and so on. It's a lot of work.

The other thing a mentor may help you with if you're lucky is to provide some insight into how to live life as a good academic – a good life in the ethical sense. I had two key mentors in my department. These, of course, were David Harvey, who taught me a lot about being a politically committed, rigorous academic, and Reds Wolman, who taught me about openness and the joys of discovery. Neither of them ever urged caution. They were always in favour of taking on new issues that you don't know anything about or taking on the amorphous, probably impossible tasks and just seeing how it goes. That was a big help and is part of how I was able to throw overboard twenty years of accumulated intellectual capital and start over in an entirely new field.

From corporate strategy to social and environmental history

The new project involves the social and environmental history of gold and gold mining. I'm working on it with a colleague in public health at Hopkins, Ellen Silbergeld. Reds introduced us because he thought we could do something interesting together although, characteristically, I'm sure he had no particular idea what that might be. We worked it out in long walks with our dogs, Sasha and Docker.

The project, in brief, tries to analyze why gold has social value and how that is connected with the distribution and the magnitude of the social and environmental costs of gold mining through the ages. Gold is such a peculiar substance. It has no real material uses – you can't cook or hunt or farm or shelter yourself with it. At the same time, it doesn't corrode, so it never goes away. Most of the gold that has been mined in history is still available. So we do a tremendous amount of damage to get it out of the ground and then mostly what we do is dig another large hole and put it back in the ground for safekeeping. The whole process raises in striking form questions about the relationship between social and environmental values in different social formations and geographical circumstances.

It turns out the project also turns distinctively on questions of power and knowledge, although as usual I didn't anticipate this at the outset. Mining in general and gold mining in particular impose a range of severe social and environmental costs. For gold mining, these include the hyper-exploitation of slave and free wage labour and crushingly arduous physical work. Mining gold has also involved the liberal distribution of mercury in the environment for a thousand years, supplemented, but not entirely replaced, by cyanide over the last century or so. Exposure to environmental toxins spreads far beyond the mining operations and imposes severe health impacts on populations that have nothing to do with mining.

The interesting thing is that these costs have been accurately assessed and commented upon since antiquity. There is no shortage of knowledge about the social and environmental costs of gold mining, but we haven't been able to use this knowledge to curb them to any great degree, despite thousands of years of experience. The reason for this, we believe, is that the one consistent social use of gold over the millennia has been connected with the amassing, display and use of social power. Until fairly recently historically, there has been such a close ruling-class monopoly on the substance that gold and social power have been virtually isomorphic. The users of gold were unaffected and untroubled by the costs which were imposed on populations and places that they had nothing to do with. Even when the information was brought vividly and powerfully to their direct attention, and even when they were Christian enough to worry about the fate of their souls, the rulers' greed for the substance outweighed their moral qualms. So the knowledge was, in effect, cancelled out by the power relations manifested in the metal.

This change in research orientation derived from several sources. One was living in a department with environmental engineers who were never going to see the value of geography if it didn't speak directly (and very slowly and distinctly) to them. This effect was amplified by my growing impression that the engineering *students* as opposed to the faculty were, in fact, very interested in what geography had to offer. So I wanted to reorient my teaching in a way that was more helpful for them. Another factor, of course, was living in a department with Reds Wolman, who is an enthusiastic and inventive interdisciplinarian and who truly is committed to advancing work that explains how we humans are connected to the natural environment and what that means. In sum, I felt a growing need to address environmental questions directly in my teaching and research.

Also, as an economic geographer, I had increasingly the sense that history matters more than we've been able to acknowledge in our work. We tend to use history as background – it provides the context and marks the boundaries of more or less coherent periods. But we're not really growing the history through our work. We are being historicist, in the sense that one thing follows another, but not fundamentally committed to the historical part of historical materialism, even those of us who still take the term seriously (Harvey 2001; Soja 1989).

This is not necessarily a call to the archives. Geographers have a tremendous amount to offer through original synthetic thinking over the *longue durée* and the *grande espace*. In part, the work of historians, like the material gathered from corporate interviews, is our data and not the answer to our questions. We need, of course, to reckon seriously with their interpretations, but we should not simply abstract their conclusions into our own material as a way of foreclosing argument about a question rather than opening it up.

Interlude: academic activism

I can't make any generalizations about whether one can normally do activist work while untenured and expect to survive. I had a lot of protection from my mentors. They wouldn't have allowed me to be turfed out simply because of my politics. It's also the case that one of Hopkins' genuine virtues is that it takes academic freedom seriously (and it is a private university so it doesn't have a deranged state legislature to answer to).

Activism also takes time that would ordinarily come out of professional work and/or family life. Since I wasn't raising a family, I undoubtedly had more degrees of freedom on that side as well.

That said, for a leftist, becoming an academic can feel like a kind of surrender. Except for very particular historical conjunctures – which have, so far, occurred twice in my lifetime – it is definitely not life on the barricades or with The People. Quite the contrary: all the universities I have spent time in are intimately part of the imperialist–militarist project that I have been railing against. It is, on the other hand, life in a place where you can earn a living and not be absolutely crushed by the system, but the price you pay for this relative freedom and sanity is not trivial. You have volunteered to remove yourself to a very rarified and select atmosphere and surround yourself with an equally rarified and select group of people where you can do work that you find immensely satisfying in many ways. You are taking a more comfortable way than is available to other activists or to the people you imagine yourself fighting for.

For me, anyway, it was both possible and necessary to do work on the side that felt more like political work and that was unrelated to my professional career (although it has been reasonably connected to my professional *work*.) In Baltimore and at Hopkins, one obvious issue for practically the whole time I've been there has involved the living wage. And really I haven't done much more than support the students and organizations that really have been on the barricades about the living wage in the city and at the university. But given my social position, it was meaningful and productive support. This doesn't get me off my own hook about whether being an academic is morally justified. But it is better than nothing.

Becoming a geographer

I didn't start out as a geographer and I want to comment briefly on why I have found a home in the discipline. When I was a kid, reading was a vital refuge from a reasonably stressful family life. But I was the only real reader in the family and we simply didn't have that many books in the house. However, being lately middle class, we did have an encyclopaedia, and so I read that from start to finish. I also had a much-cherished great aunt who did have a big library, but it was largely accumulated through the Book of the Month Club and so was quite random. As a result, I think I am probably naturally intellectually undisciplined, which in polite company I call being interdisciplinary. Because of this, because of my anti-establishment politics, because of many things, I find being in a discipline with firm precepts about what it thinks about and how it thinks impossible. I need the big, capacious disciplines that can go anywhere, that are methodologically heterodox and ontologically flexible. This simply feels right to me. I also think it *is* the right way to gain an understanding of the world. Geography in my view is close to being the perfect discipline and I don't understand why more people don't see that.

In a way, however, I do understand. My beloved colleague Reds has described the discipline as being more of an aspiration than an accomplished fact. It wants to do things such as truly integrating the physical and the social that are inherently difficult and perhaps impossible. We keep not quite getting there and yet we keep on trying. I think Reds is right and that this aspirational quality is a source of real strength at the same time that it makes it hard to be clear to outsiders and sometimes to ourselves about what we're up to. I think it is useful for us to check in with each other about what it is we think we're doing from time to time, which I take to be the purpose of this book. But I would resist any attempt to discipline the discipline. We owe it to ourselves and to others (though they do not know it) to defend this undisciplined space.

NOTES

This chapter is an expanded version of a talk given at the first Summer Institute in Economic Geography organized by Kris Olds, Jamie Peck and Adam Tickell. The audience was a group of advanced graduate students and junior faculty in economic geography. Though the present volume has a broader intended audience, in writing this I have kept my focus on what might be most helpful to people coming up in the field.

1 My thanks, I think, to Peter Jelavich for raising this point.

2 SMOKE AND MIRRORS: AN ETHNOGRAPHY OF THE STATE

Alison Mountz

In 1999, the federal government of Canada intercepted four boats carrying migrants smuggled from Fujian, China to the west coast of British Columbia (BC). Most made refugee claims and were detained and ultimately deported. This episode brought to public debate Canadian refugee and border enforcement policies and the bureaucrats who administer them. My chapter is about those bureaucrats. Despite standing on the inside looking outward as they enact immigration policy, civil servants often, ironically, express feelings of powerlessness, particularly about human smuggling. This episode, therefore, provided an ideal opportunity both to see the state and to see like a state (J. Scott 1998). The chapter is an ethnography of state practices.

This paper argues that ethnographies of the state, while challenging to undertake, are an effective tool to study states as daily entities. To develop this argument, the paper draws on research conducted at Citizenship and Immigration Canada (CIC), the lead department responsible for the management of transnational migration. The research examined the response to the four boats as a case study. More generally, the paper explores the contours of the emerging cross-disciplinary genre, 'ethnography of the state'. I am particularly interested in the methodological challenges of undertaking research within a bureaucracy and in what the very research process reveals about 'the state'.

I must have written this paragraph above in some incarnation at least twenty times. It tends to come out the same way, mocking me on the page, reminding me that I have now mastered the language that once alienated me.

As a sociology major at university, I tired quickly of academic prose. I would skim the theoretical arguments of the text and skip ahead to the best part: the interview responses and researcher's fieldnotes. These were typically printed in small font, indented, and sandwiched between loftier words that obscured ideas in long sentences. Authors literally minimized these quotes, devoting as little room on the page as possible to what I thought were treasures: the vocabulary, syntax, emotions, and daily struggles of a community or event under study. I wondered what if this hierarchy was reversed.

Anthropologist Ruth Behar (1996) has been moving the small print into large print. She calls this 'emotional anthropology' and argues that it is the only kind of anthropology that matters. Sheltered still by the cloak of objective social science, I cannot quite bring myself to write Behar's emotional anthropology ... nor can I seem to put her books down.

Geographic understandings of the state would benefit from her approach. The narratives that I collected in the field contradicted popular stereotypes of the mundane, robotic nature of bureaucratic work. Interviews were infused with emotions surrounding the controversial and contingent response to human smuggling. I know only too well that the researcher's emotions also came into play. Their inclusion adds a key dimension to conceptual understandings of state practices. Like Herbert's (1997) experiences on patrol with police officers, the emotional encounters of participant-observation underscore the degree to which human agency drives the uneven implementation of public policy and the law.

But how would the emotion of my own ethnography of the state translate from practice in the field to text on the page? This chapter incorporates autobiographical experiences, including emotion, to highlight the degree to which performance and performativity are at work as academic and civil servants constitute one another through ethnography.

> I am interested in what the research process reveals about the bureaucracy, the social process that helps to constitute knowledge. Using excerpts from fieldnotes, this paper conceptualizes fieldwork as performative of the state (e.g., Goffman 1959; Pratt 2000; Wolf 1992). This approach, I argue, highlights the contingent, contextual, dynamic, and performative practices of the state.

One morning I arrived at my desk at CIC to find yet another hot pink Post-it note stuck to the monitor of the computer that had been carefully installed with an outside telephone line to dial up the internet, but studiously without access to the internal network of the Department. I knew, even before I had read it, what the note written hastily in coloured ink would say: yet another immediate visit was required to the office of the person currently managing human resources, the mid-level civil servant in charge of monitoring my access to materials and more generally my whereabouts at CIC's Regional Headquarters (RHQ) in Vancouver. I had come to know this woman in my mind as 'The Watchdog'.

And so it was that I entered the Watchdog's office that morning in August 2000 for what turned out to be a bizarre exchange. She informed me that I was not to enter 'the secret room' under any circumstances. Confused and concerned that she might mean the room behind the fortified wall where the intelligence analysts from whom I had been learning a lot were located, I asked, 'Do you mean the room where Jim is?'[1] She responded, 'I don't know. I can't tell you. You're just not to enter the secret room.'

Days later, another employee insisted that there was nothing in that room that I could not see; that he had seen everything, and that there were no secrets. A few months after that, someone higher up in the Department revisited the issue — with no prompting on my part — and argued that there were in fact secrets filed away that even he was not allowed to know.

I was perplexed, never having expressed an interest in secrets or the secret room. I had, however, discovered a pattern that would repeat itself in the months to come: the contradictory assertions that valuable pieces of information existed and must be hidden, and that there was nothing to hide. This lesson repeated itself even in the form of others' reactions to me as a researcher. Some bureaucrats perceived that I too had something to hide and was faking an injury to elicit sympathy and support for my project.

Because of a knee injury I walked with the assistance of metal crutches during that first month of fieldwork, so it was not difficult for anyone to monitor my whereabouts; especially since the office was so quiet — one employee referred to it as 'the morgue'. One morning as I wandered by the printer, an older female administrative assistant remarked, 'When this is all over, Alison, we're going to find out that you were faking it the whole time.' It was the 'when this is all over' part that jarred me most.

As Herbert argues, participant–observation offers the opportunity to observe not only what people *say*, but what they *do* (2000). This characteristic of ethnography is essential to ethnographies of the state. Multiple agendas are at play beneath the surface, yet institutions must maintain an outward façade of coherence. One civil servant once described his work as a member of an office staff running madly to maintain a façade. The researcher who is able to step behind the public façade to study the daily work of bureaucrats may stand at odds with the project of communicating a coherent narrative.

> Approximately one year after the first boat arrival, I began ethnographic research with CIC, the Department responsible for the interception, processing, and detention of migrants. The methodology was designed to embody and thus demystify the state. I was interested in how individuals within the Department made sense of the boat arrivals and of their own role in responding. I gathered narratives of the 1999 response to human smuggling in BC through participant-observation, semi-structured interviews with civil servants and other institutional actors involved in the response, archival research, and media analysis.

It was June, two months before my encounter with the Watchdog, when my PhD adviser (David Ley) and I took the bus downtown. We had a meeting with the director of the CIC regional offices to see if this project might be possible. I privately called this director 'The Big Cahuna'. When we arrived the 'Cahuna' was out for a walk, having forgotten our meeting. His assistant seated us at a round table in his posh corner office overlooking a busy commercial intersection. After she left, David turned to me and said, 'This is the part of the movie where we search the drawers and ransack the office!' We laughed.

The Cahuna finally arrived, and I liked him immediately. He was frustrated that day because after so much hard work the new Immigration Act tabled in Parliament had been postponed until next year's sessions. He had gone for a long walk to cool off. The Cahuna turned out to be a progressive thinker who had studied geography and took an interest in my research and what he labelled the potentially 'objective' view of an outsider on immigration matters. He decided to take a risk that day, and I owe much of the eventual success of the project to him.

So, on and off for several months, I participated in the daily exchanges of office life at Regional Headquarters, without actually doing any immigration-related work. I occupied a desk in the office, reviewed documents, conducted interviews, and 'hung out' with employees. This served as a base of operation where I attempted to interview everyone involved in the 1999 response to human smuggling in some capacity, from frontline officers who boarded the boats to officials located higher up in the administration of the Department in Ottawa.[2]

Interviews lasted an hour or so, but I met with some individuals repeatedly for extended periods. On occasion, I tape-recorded interviews, but found that this inhibited the exchange significantly. Sometimes I offered to stop writing altogether in exchange for more candid discussion. Not all that had come to pass would be recorded on paper.

> Because of the importance placed on protecting information and producing a set of clean public narratives, it is challenging for social scientists to access the state and other powerful institutions ethnographically. This research, like all field projects, carried with it a host of ethical quandaries. These included the desire to protect those who participated in the study as well as the privacy of their clients. The bureaucracy is designed to protect employees, clients, and information.

On that first day, after completing a security check, I was given a friendly tour and round of introductions in the office. I was then shown to my desk, based in the Information Technology (IT) Department. IT employees worked for a while to set up internet access for me. In an e-mail to my adviser during that first week I wrote:

> *The state continues to be a friendly entity. They've set me up in a cubicle with a big desk, computer, phone. They have granted me 24-hour access to the building. Part of me wonders if they want me to identify as being a part of the state, the friendly state.* (4 August 2000)

The Cahuna's desire to be open nonetheless conflicted with security issues. IT employees tried, for example, to put me on an e-mail listserve for the staff, but this never came to fruition.

As with any research project, things were awkward in the beginning as I tried to learn my place and as others tried to figure out what I was doing there:

> *It's just so strange for me to be in this office, strange for me and for them I know. It's always like this in the beginning of fieldwork though so I'm hoping to build rapport and trust and get to know people. It's draining to be here and to figure out my position.* (14 August 2000)

During that second week of fieldwork, I met with employees working on human smuggling and presented my research questions, objectives, and methods. The goal was to introduce the project and garner feedback from potential participants. We went over the consent form and the ways in which I would protect both the information I gathered as well as the identities of respondents. The meeting provided an opportunity to meet a number of people. Soon after that, though, things began to unravel. I learned that the first director, who had agreed to let me do research, was to be replaced. I was worried. I met the new director on his introductory tour.

> *Liz introduced me as 'a graduate student investigating us' whom they decided to allow in to do research.* (16 August 2000)

My problems began. One individual who attended the meeting where I presented my research approached the Department of Justice (DOJ) about the project. CIC then commissioned the DOJ to write a legal opinion. It was not favourable, precipitating a series of negotiations with DOJ lawyers. They placed restrictions on my work. Hence the Watchdog.

DOJ lawyers put the following conditions on my research: that I interview people with a lawyer present, that legal counsel review the transcripts following the interviews and that CIC own and destroy data. Many of these requirements, however, conflicted with the conditions imposed on me by the UBC Behavioural Research Ethics Board, such as protecting the anonymity of interviewees. These negotiations demonstrated the differences between the bureaucratic mechanisms designed to protect information and proximity required to conduct research. Lawyers treated me as a potential journalist engaged in accumulating damaging information rather than as an academic researcher.

Things grew even more complicated. My 'advocates' left information on my desk for me to read, while my 'opponents' monitored what was there and chastised colleagues for sharing information. This all left me in the dark and frankly, a little paranoid.

*Now I have some of everybody's stuff on my desk. And it's all out on my desk,
and I don't know if it's the best thing. That desk is too public. ... How could
I make it seem like more of a private workspace?* (22 August 2000)

Meanwhile, the Watchdog monitored my every move.

*I came out into the hallway. ... Sally was getting on the elevator with Liz. She
said, 'Oh, Alison. Should we hold the elevator for you, are you coming down?'
I said, 'No, I'm just going to the bathroom.' And she said, 'You sure came the
long way from your desk.'*

The state treated me as an object of its daily practices at the same time that
I attempted to turn into text the daily practices of the state. Civil servants
and academics found themselves both subjects and objects of analysis.

Eventually, I was told that I was neither allowed to read anything, nor to
talk with anyone. When I asked directly about interviewing people, I was
informed I could not ask for 'personal information', which meant anything
anyone said. The lawyers worried about open-ended questions and feared
someone might say something damaging. It was eventually decided that I
would read only those materials released to the public until the lawyers wrote
a contract detailing the conditions under which I could undertake research.

It became clear that the lawyers were not interested in research but in
avoiding the possibility of litigation. In this sense, they performed the
power of the state with vigour, protecting information that they had not
seen but believed to be present and in need of guarding.

On one of my last days in the office, I wrote:

*I've got lawyers in the Department of Justice after me and people following
me around the office telling me I can't see secret things or be in secret places.
There are accusatory letters going around. It's the stuff of movies, or better
yet, the stuff of bureaucracies. And in the middle of all these power struggles
is little old me, along for the ride.* (31 August 2000)

On 5 September, my last day during this initial phase at CIC, I faxed a
letter to everyone involved asking for resolution. A couple of weeks later, I
met with some contacts at a coffee shop outside the office and learned that
there were orders issued for my removal from the office the next time I
entered. This coffee date, one of many, prompted contemplation of the via-
bility of an ethnography of the state that took place beyond the office, on the
fringes of the bureaucracy, much like what Philippe Bourgois (1996) calls a
school ethnography, comprised of narratives of those who had dropped out
of school while they were hanging out on school playgrounds after hours
to recount tales from their educational pasts. Would an ethnography consist-
ing entirely of coffee shop meetings, without participant-observation in the

office be possible? Meanwhile, the Department officially withdrew support without officially telling me.

Things moved slowly. I was unable to access anyone officially in CIC in September and October, making telephone calls that were never returned. Instead of writing a paper based on my research, I wrote a paper about the difficulties of doing any research at CIC at all. As a direct result of that paper, I was invited to meet with the new director and a lawyer from the DOJ, to work out a return. We established a new *modus operandi*, and soon after, I met with Department officials to negotiate a new agreement. The epiphany during this meeting was that when the study was over, the Department could simply dismiss my research findings as mere 'opinions' and be free of blame.

I resumed research in January 2001 when the Department did another turnaround. The director circulated a memo to employees welcoming me to the office and encouraging support and participation (albeit with reminders of rights and responsibilities of civil servants). I was given copies of files released to the public and granted access to review in-house files that pertained to the response to smuggling. On occasion, someone would check through what I was reviewing but, in all, the Department was supportive of my research and granted permission to conduct interviews in the 'translocalities', Akhil Gupta's (1995) name for the multiple locations of the state.

I spent the bulk of my time conducting interviews at CIC and other institutions in Vancouver, but also made two trips each to Victoria and Ottawa, and one trip to Hong Kong. I was thus able to follow the networks of people who worked on the issue of human smuggling from a variety of locations. In Hong Kong, helpful contacts in CIC introduced me to their colleagues from other consulates with whom they shared information through working groups and informal networks. The performance of the state had shifted dramatically, and the success of my project depended now largely on personal relationships among co-workers.

Increasingly, social scientists are extending Erving Goffman's (1959) depiction of human behavior and fieldwork performance (Katz 1992; Pratt 2000). Goffman suggested that as people interact, they act out roles with a desire to make particular impressions that may or may not be perceived as intended. More recently, poststructural and postmodern theorists have conceptualized identity as fluid and incomplete, forever shifting and constituted contextually. Judith Butler's (1993) notion of performativity relies on the idea of citational iterations. This concept differs from Goffman's view of the intentionality of human agency and focuses instead on the ways that identity is assigned meaning through discourse.

Civil servants performed the state differentially. Some people interviewed expressed frustration with the amount of power wielded by the communications branch, those charged with presenting policy to the public. I had initially thought that communications employees might be the most willing to co-operate. It was the opposite. They reacted warily to an individual stepping behind the public image. Intelligence officers, on the other hand, whom I initially thought might be guarded, turned out to be more open, perhaps because their own work corresponds most closely with the research process. This juxtaposition of standpoints pointed to the importance of public image for the Department. Ethnographic study revealed daily practices concerned with maintaining a public image of being in control. Respondents began every day by reading news clippings of media stories that mentioned CIC. They also reminded me frequently that there was no room for mistakes in the public eye.

Just as civil servants performed the state in contradictory ways, I enacted a performance of my own at RHQ: that of academic researcher. But my presence made people uncomfortable. Implicit in the joke-that's-not-really-a-joke at the printer was the assertion that I had been faking an injury to gain access to the bureaucracy. A few months later when I finally sat down with someone who managed my access (albeit who had never met me), he said that he was surprised to find that I was a nice and reasonable person. He had heard that I was 'a real bulldog'.

It is easier to detail these experiences without mentioning how emotionally challenging they were. Many mornings I wanted to stay at home rather than risk these difficult encounters. Fieldwork is exciting, but it can drain energy and self-confidence. Sometimes I would buy and slowly sip a coffee at the shop downstairs (aptly called 'Death by Chocolate'), writing in my field-book to steel myself for the day ahead. That August, I craved the one day a week that I would return to the university campus and fit in. I was conflicted between a desire to retreat from the world in order to write, and a desire to interact with the world in order to write.

During the period of the most difficult negotiations, I began dreaming about the situation. The dreams involved me misrepresenting myself in order to invade others' privacy. They also turned on the confusing geographical boundaries between places where I felt that I belonged and where my presence made people uncomfortable. Performance and performativity of state and research thus came into conflict, and I began to analyze my own performance in an unexpected and unintended inversion of ethnography.

We know that something is afoot with the state. Scholars have been returning to the state with increasing frequency (e.g., Brenner et al. 2003) to argue that the exercise of sovereignty is shifting (e.g., Hardt and Negri 2000), that states are at once more and less powerful, that power is at once more concentrated and yet more dispersed, informal, and penetrating.

Secrets and policies were performed differentially, and this realization went to the heart of my inquiry about how civil servants saw and responded to human smuggling. I spent many hours meeting with a mid-level bureaucrat who was knowledgeable about the Department's research on organized crime. Subsequently, he told me not to believe any of the narratives circulated by his colleagues about organized crime. In his assessment the declaration that transnational organized crime had facilitated this smuggling enterprise was an easy narrative promoted by leaders and communications employees alike. But there had been no names named and little substance behind the story. As the earnest researcher, I recorded this information only to be reminded that nothing was as it seemed.

This respondent knew what feminist scholars have argued for years: there is no underlying truth to be discovered in interviews, only a series of narratives that people tell, performances offered at distinct moments for distinct reasons (Rose 1997). These are the clues to understanding the state through ethnography. When and where were different narratives and conspiratorial tales called upon and why?

I came to understand civil servant performances as occupying the gap between a state in control and the more frenetic reality of daily work that is dynamic and uncertain. The generally well-oiled and well-resourced performance of communications employees is designed to counter the performative iterations of public discourse wherein 'the state' appears powerless and disorganized.

> Ethnographies of the state are uniquely able to illustrate the fluid, diverse array of subjectivities and discourses through which states are constituted (e.g., Gupta 1995). Ethnographic data illuminate the very life-like, uneven aspects of state practices. The state is as much an idea as it is a material reality; and the material realities that structure and emerge from state policies are dependent, in part, on whose idea of the state is being performed in any one particular place and time. As such, institutions embody inherently contradictory sets of ideas, and ethnographic data demonstrate the performative nature of human behaviour.

I was jittery the day that I went back to CIC to present my work, my final performance. I was happy to be taken on a small tour of the office where I learned that the old carpet had been replaced. The new one's design involved a swirl like Nike's 'swoosh' whose spiral could be traced through the office yet ultimately led nowhere. It offered a fitting backdrop to life in the bureaucracy.

Prior to my presentation, I had an indication of how my work was being received. I had passed chapters along to my closest informant with a litany of caveats and conditions surrounding the text. He laughed, took the chapters home, devoured them, reported that 'it almost read like a novel', and placed a pre-order for two bound copies for the office. He then brought the chapters into work and passed them around. I heard the story of three people sitting around in an office one day discussing the text and trying to identify colleagues interviewed and quoted.

When I turned off the projector, having shown my last overhead, I braced myself. The lengthy discussion that ensued was far friendlier and funnier than I had anticipated. Some people discussed how to distance themselves from the media during an operational crisis, while others still worked to determine who had said what in the quotes provided.

The Big Cahuna died suddenly before I completed my dissertation and had an opportunity to share it with him. It was profoundly sad for me and for the hundreds of people with whom he had worked over the years. This entire experience would have been complete only if he could have been there on the day that I presented my work, back in the boardroom where he had created the space of possibility for the project. Instead, he reads over my shoulder as I write.

> This paper has advocated empirically grounded ethnographies of the state, uniquely situated to detail the performativity of identities that constitute day-to-day state practices. Social scientists are theorizing the changing nature of power and of states. We hear that states are more diffuse, dispersed, and transnational; that they have devolved and dissolved. But what do these theoretical assertions mean? What do they look like on the ground with changing practices of border enforcement, for example? How are such changes embodied in the daily work of civil servants?

Writing, it turns out, is another performance of the state and the academic. Some readers might think this chapter is only about me, not the state. But I would disagree. The research was about me *and* the state, and the fleeting affair that we had. My own dreams and entanglements were a mere glimpse into bizarre twists and turns of public policy sustained during the course of a civil service career.

I relayed these field experiences in order to link ethnography to different understandings of the state. This methodology corresponds with my conceptualization of state practices as networks of everyday, embodied practices and relationships. I write always against those interpretations that take public policy at face value. Something that quacks like a duck may not necessarily be a duck. I enjoy the story that leaves interpretation between the lines. If we are to understand the state as a dynamic, contextual set of practices, networks, and performances propelled by engagements between human agency and public discourse, then 'the state' on any given day owes as much to the observer as to the observed. This ethnography is therefore at once autobiographical and about the state.

Critical human geography brings to ethnographies of the state the impulse to locate and analyze the spatialities and social relations of state practices and power in multiple locations. These include everything from the locations where interdiction and detention take place, to who is in on conference calls, and to the ways that the micro-geography of office floor plans relay spatialized power relations. But the places where civil servants and researchers connect and disconnect also reveals a micro-geography that is heavily coded. Ethnographers of the state can never quite be 'in place', for they face the challenge of documenting life in a place whose narratives about itself may be distant from the realities of daily practice. Ethnographic research locates, names, and interprets the power of institutions and people within them. These institutions, in turn, may craft the researcher performatively into their own object, naming, locating, and interpreting her motives. Academic and state therefore serve as both 'object' and 'subject' of ethnographic research.

Acknowledgements

I thank Jennifer Hyndman and Helen Watkins for comments; Trevor Barnes for editorial work; David Ley, Gerry Pratt, Dan Hiebert, and Vicky Lawson for encouragement; and many people at institutions in Canada without whom research would not have been possible.

NOTES

1 All names are pseudonyms.
2 Various institutional actors within and beyond CIC took place in the response to the boat arrivals, supporting the notion that 'the state' is a heterogeneous category with permeable boundaries (Mountz 2003). In addition to federal and provincial bureaucrats, I interviewed employees of law firms, media institutions, non-profit organizations, activist movements, and supra-state bodies.

3 NATURE TALKS BACK: STUDYING THE ECONOMIC LIFE OF THINGS

Paul Robbins

My first encounter with the mesquite tree (*Prosopis juliflora*) was on a desert road in India, under a blazing sun, tens of kilometres from the nearest settlement. When the front tyre of my motorcycle went suddenly flat and the bike skidded and dumped itself and its rider in the dunes by the side of the road, I did not immediately know the cause of my misfortune: a twig from a thorny invasive tree that had made itself at home halfway across the globe from its original habitat in North America.

Subsequently there were other encounters, but it was only as I tried to explain and understand the political and economic history of the Marwar region of Rajasthan some years later that the full import of the gnarled branch began to dawn on me. The long thorn that damaged my tyre came from a species that travelled tens of thousands of miles to wedge itself underneath my wheels, seizing opportunities, making friends and enemies over more than a century, and participating in the total transformation of the region's rural economy.

It is common to hear researchers lament the lack of engagement with the non-human elements of geographic reality. To do so, however, presents several puzzling questions. How do non-human actors set the terms of political economic reality? How are they remade by the systems of power in which they are enmeshed? How do they act to remake us? And how would we know?

This chapter seeks to answer these questions by briefly reviewing three discreet cases of non-human objects acting in the world and producing different economic geographical effects: mesquite, an invasive exotic tree in northwest India; Chronic Wasting Disease (CWD) found in wild cervids (e.g., elk, deer, and caribou), and lawn grass, one of the dominant biotic land covers of North America. In each case, I attempt to show both the importance of these objects in economic geographic reality, but also the problems they present in undertaking research. I show how my own research techniques have evolved by conducting a series of conversations with and about such objects, illustrating how they participate in the creation of economies and polities, and explaining how such participation forces difficult methodological choices on the investigator.

Though *Prosopis juliflora*, CWD, and lawn grasses, have all personally confronted me at different times, the lessons learned in addressing them, I argue, extend to the whole population of non-human objects around us, from tractors, computers and ATMs to houseplants, mosquitoes and viruses. All of these non-human objects 'talk back' constantly. Learning to hear them is the ongoing methodological problem.

The chapter is divided into three sections. First, I argue that mesquite-covered landscapes are both a product of colonial and postcolonial economic conditions, and a driver and participant in ecological change. Departing from traditional landscape history that deploys a combination of 'natural' and 'social' forces to explain environmental outcomes, I argue the mesquite tree is best treated as a *quasi-object* (following Latour 1993), linking multiple actors.

Second, my research on wildlife privatization in Montana forced me to understand the incidental flow of non-human materials as a product of the economy. In particular, I approached elk, people, and proteins joined together through Montana's wildlife enclosures as a collective or *hybrid agency* (following Mitchell 2002). This concept revealed the way such objects behaved, the opportunities that they presented for themselves, as well as the prospects they created for other actors, frequently unconnected to the best-laid plans of privatization and capitalism.

The final section reviews my encounter with the turfgrass lawn. Though I began this research with much the same method and focus of previous work on the topic, trying to explain the cultural economic forces producing this singular form of nature, I was driven in a reverse direction, and forced to reflect on how grasses act on the ideological subjectivity of householders. Following the work of Louis Althusser, I suggest that turfgrass lawns participate in complex economies by *interpellating* (or 'hailing') people into self-recognition (following Althusser 1971).

Trees are political agents: researching quasi-objects

Consider how economic geographers research trees. Generally, such research either explores or explains aggregate tree cover (as 'forest') as a function or indicator of social and economic change. In conventional land cover research, forest cover is seen as an indicator or responding variable, a *natural* artefact of *social* conditions. In more critical work, the forest or trees are seen as *natural* objects of *social* struggle, resources or sites of production or reproduction that set bureaucrats, powerful interests, and marginal groups against one another in a political geography of resource control (Peluso 1992).

Each approach brings with it a reliable and well established set of methodologies. In the first case, changes in forest cover extent or morphology are measured remotely with air photos or satellite imagery. Areas of change are correlated with other local factors, population growth, changes in land tenure, urbanization, or distance to access points and roads (Turner et al. 2001). In the latter case, intensive ethnography might be employed to explore how people use forests, how claims to control and ownership of forests are contested, and how competing claims about forests and forest change are used for political ends (Hayter 2003). Such approaches provide a research blueprint for economic geography, asking relevant questions and producing meaningful answers.

My own encounter with the economic geography of trees, however, failed to follow this tidy course. Recovering from my motorcycle accident, I continued to examine how strong state authorities carved up landscapes into specialized uses, carefully producing a natural geographic landscape to mimic a geometric social imaginary. Such a process, which Scott describes in his book *Seeing Like a State* (1998), is predictable, because the modern state and capitalist economy must 'simplify' the landscape in order to produce and control it. The result, he posits, is a change in the character of forest cover and its ecological diversity.

To approach the problem this way requires measuring landscape change over time, and mapping its shape and form using satellite imagery, air photography or other means. It further requires showing how economic forces – taking either the form of agricultural markets or political economy struggles – are at work in forest transformation, by interviewing producers, wood extractors, farmers, and forest managers, while examining agricultural and timber markets and commodity prices. The external pattern of nature, visible in landscape change on the ground, is then understood as an expression of the latent order of the economy and the state.

And yet evidence of chaos, rather than order, followed me wherever I went. The mesquite tree littered the landscape in apparently random patches, violating the clean boundaries of the planning process. Both in field-based explorations and in time-series analysis of remotely sensed images, it was there. Found where it was desired and where it was not, the mesquite created forest-like conditions in non-forest places, but was also deteriorating the integrity of stands of trees that had historically been considered forests.

These patches might be conceptualized as random noise occluding the underlying economic order. In this view, the mesquite is a 'contingent' outcome, produced by idiosyncratic local conditions that in no way impinged on the otherwise rational relation between economy and landscape. Such an interpretation is tempting. The alternative, that trees represent a pattern, though not one necessarily following from any essential or even fully human process, would require a transformation of research questions and methodologies.

I became increasingly convinced, however, that the landscape pattern of the mesquite was by no means random or contingent. The presence of this ubiquitous tree may itself be a product of the very landscape simplification that it defied. Or perhaps more disturbingly, the tree might be driving social and economic change. As Bruno Latour has suggested, the best way forward in explanation may be to abjure using society to explain nature or vice versa, but instead to start with socio-natural phenomena: quasi-objects. Using these as a starting point, research takes the non-human seriously without evacuating explanation of the political and economic. The technique, therefore, is to identify apparently anomalous socio-natural objects that destabilize the discreteness of nature and society (quasi-objects) and trace their biographies to discern what makes them so troublesome.

To explore this possibility meant turning my own thinking inside out and practising a very different kind of geography, one that neither views the landscape as an accreted product of cultural and natural influences (Sauer 1965) nor as an environmental tableau on which political struggles are fought out (Bryant 1992). More practically, it meant understanding better the tree itself, its ecological characteristics, its social position, its friends and enemies, and its economic biography. This implied: poring over botanical forestry data about the species stored in Indian state research records; sifting through archives (located thousands of miles away in the UK) for specific mention of the tree's historical role in the region; and walking with countless people, including foresters, herders, and farmers, to understand their relationship to mesquite.

As it turned out, the tree has long had a rich economic and political life. Transported to the subcontinent from the Americas by colonial officials more than a century before, mesquite was selected precisely because of its limited value in subsistence, especially grazing. Extolled as a survivor, largely due to the inedibility of its leaves by livestock, including goats, the tree made an excellent selection for foresters whose ideology was to 'green' the landscape at all costs.

But as the tree began to establish itself, it came to influence the very experts who sought it out in the first place. Owing to its lateral canopy, it tended to increase the incentive for remotely sensed census techniques and canopy measures in forestry. Because it makes a fuel source, even while being largely inadequate in many other regards, it tended to collude towards the prominence of fuelwood as a motive for forestry. The tree had found colonial allies, who served its interests and whose interests were in turn served.

These same complex relationships extend into the present, with farmers, herders, and foresters interacting differently with the tree, struggling against it in some areas, aiding its expansion, either intentionally or inadvertently in others. The species in turn continues to influence the economic practices of local players, their use of land, and their views of nature, even while the biogeography of its advances and retreats are in turn shaped by these same players (Robbins 2001). Nature is 'talking back' through the tree, but the very category of nature itself, aloof, distinct, and discernable, begins to dissolve.

Prions connect humans and non-humans: assessing hybrid agency

The lessons of this work had hardly had time to settle into my mind when my attention shifted to an apparently unrelated problem in environmental management – elk populations, wolf reintroduction, and the hunting economy in the USA. The focus of the project was to create a viable model of elk, as they moved across the Northern Yellowstone Range, in and out of America's most politicized national park, encountering social forces that regulated their population.

If the model were to run correctly it would have to incorporate not only ongoing changes in human land uses in the region, like the conversion of ranches to more dense subdivisions, but also shifts in policy, including enclosures and rules for controlling and hunting the animals. In each case, there has been upheaval over the last decade as demographic and economic shifts have changed rural Montana from a place of primary economic production to a site of consumption, with second homes, amenity buyers, and tourists replacing ranches and farms.

Again, the intuitive approach to the problem for an economic geographer would be to show how such social and economic shifts in the human world 'write' themselves into the landscape with concomitant direct and indirect effects on elk populations, behaviours and adaptations. And starting from this perspective, my efforts focused on surveying different groups and providing a policy overview of the context, its political ecology, to inform various scenarios for modelling possible elk futures.

Once more, however, the unidirectional processes for which we searched were constantly muddied by the metabolism of the socio-environmental system. Specifically, interviews and policy history revealed that techniques for the control and enclosure of elk were going through a period of dramatic upheaval. Experiments in privatization included direct enclosure of animals in fenced hunting areas or in institutional enclosures created by giving transferable ownership over the right to hunt elk. But results were mixed. While licences for new and expanded 'game farming' facilities – physical enclosures – rose tenfold between 1993 and 1996, this was followed by a dramatic decline, with the number of new facilities dropping to nearly zero by 2001. And though there were annual efforts on the part of legislators to pass laws legalizing licensing practices to increase the effective ownership of elk on private land, they faltered, with the last efforts in early 2001 ending in easy defeat in the state legislature.

The failure of this revolution to remake the relationship between wildlife and property might first be posited as rooted in economic logic. That is, emerging opposition to the enclosure of wildlife as an economic good might be understood to reflect economic consciousness on the part of opponents, articulated as opposition to the accumulation of collective or communal property: 'they' are taking 'our' elk. Indeed, advocacy groups for traditional hunters, a community historically more economically marginal than out-of-state game hunters and large landowners, blasted these enclosure efforts as 'privatization', a powerful criticism, especially when forwarded by historically conservative constituencies (e.g., rural Montana hunters) with a strong ideological orientation to property rights.

Ongoing work began to suggest, however, that the struggle over the economic status of elk was not simply one class of people (landowners) struggling against another (hunters). Rather, other players and processes were at work that again blurred the line between humans and non-humans.

In fact, the most important factor influencing the dramatic decline and fall of elk enclosure was concern about Chronic Wasting Disease (CWD), a neurological illness infecting wild cervids. The clinical signs of CWD are obvious: infected animals are emaciated, with lowered head, droopy ears, wide stance, and excessive salivation.

The disease, as it turns out, is a transmissible spongiform encephalopathy (TSE), a fatal neurological condition that science suggests is caused by abnormal infectious proteins called prions. TSEs are thought to include bovine spongiform encephalopathy (BSE, or mad cow disease) and Creutzfeld-Jacob disease, the form of TSE fatal to humans. No record of transmission between CWD and the other TSEs has been demonstrated, although such risks are being considered (Salman 2003; Schauber and Woolf 2003). Fear surrounding the disease and its possible spread into Montana herds acted as a profound break on privatization. It became apparent that understanding the political economy of elk meant grasping the details of the disease, and the pattern and flow of the prions causing it. My own work shifted in the following months to tracing and exploring these non-human actors.

While the origins of the disease are not entirely clear, CWD was possibly transmitted to wild herds penned in the same areas as domesticated sheep carrying scrapie (a TSE), perhaps as early as the 1960s. Since that time, the disease spread through wild cervid herds, with massive infection rates among wild animal populations in some areas, especially Wisconsin, where the state slaughtered tens of thousands of wild deer in an attempt to check the spread of the disease.

It is also largely agreed that when normal prions, which exist in human and animal neurological systems, come into contact with abnormal ('bent') prions (thought to be the disease agent), they are converted into abnormal ones. This 'chain reaction' rapidly expands the number of disease-producing proteins, which are insoluble in all but the strongest solvents and are highly resistant to digestion. They therefore survive for extremely long periods in soil or other non-organic contexts. More significantly, it is now well established that CWD has a much higher rate of transmission among captive herds, and is most quickly transmitted through the interstate transfer of game farm animals. In other words, prions move most rapidly through capitalized ecosystems. Disease risk thus becomes a strong argument to oppose privatization (Robbins 2004; Robbins and Luginbuhl 2005).

Such feedbacks and interlinkages appear as straightforward examples of 'biocomplexity', where social and natural systems impinge on one another. Certainly this is how they would be incorporated into the computer model, and to good effect.

But there is more at work here than simply mutual influence between 'nature' (elk and prions) and society (legislators and ranch owners). Rather, it is clear that the elk are themselves social actors, as much as legislators are ecological ones. The prions flow through the society and economy in a way that is fluid and uninterrupted. Prions, capital, landowners, legislators, and hunters form a complex 'hybrid' agency. As Mitchell observes, the agency of people is itself only a product of such connections. 'Human agency,' he observes, 'like capital, is a technical body, is something made' (Mitchell 2002: 53).

But how does one empirically grasp such a condition? What kind of methods reveal such linkages?

While requiring the normal protocols of the research, this case also necessitates interrogation of the role of prions in economic life. Because not much is known about prions, and because it is by no means a unanimous opinion that Chronic Wasting Disease is even caused by prions, putting these tiny proteins at the centre of a research project is difficult, maybe impossible. Even so, the porousness of people and elk and their biological and social connection now appear more clear to me, and this very permeability seems to be the condition that allows for political economic upheaval.

The implication is that research in economic geography might profitably begin not from people, nor indeed from abstract theoretical entities like 'the economy', and instead start from objects or more accurately quasi-objects like prions, those anomalous phenomena that blur the distinctions between the social and the natural, the human and the non-human. Is CWD natural or social? Anthropogenic or ecogenic? The intractability of such questions indicates the importance of such objects as prions as the starting place for work.

One last point: prions do not have separate causal force outside their relationship to enclosure legislation, ranchers, state veterinarians, and home builders. Agency is a characteristic of networks rather than being embodied within objects or people. One needs to trace the networks though which prions flow, and identify the nodes (people, elk, legislation) that mediate and which are influenced by the prions. Research that seeks to explain current political economies needs to trace methodologically the conjoined networks of organic and inorganic things.

Turfgrass creates subjects: exploring interpellation

If human agency is a 'technical body', it is also one experienced reflexively, in daily life. How can being enmeshed in a series of relationships with objects be reconciled with our own experience of independent agency? The answer to this question is provided in another research project of mine that has gone awry which sought to explain the presence and ubiquitousness of a typical urban landscape, the turfgrass lawn. Originally intending to demonstrate that the lawn was a product of a global chemical pesticide economy, I intended to trace the changing character of the lawn chemical industry.

The research initially followed a traditional economic geography, focused on the debt structure of the industry, its global expansion and contraction, and value added nodes along its commodity chain. Those nodes extended from large-scale chemical producers, to regional formulator companies who mix and brand pesticides and herbicides, to local communities and households, who were interviewed extensively to understand their motivations, aesthetics, and views of nature, conservation, and health.

The results of the study were at once both obvious and subtle. Wealth and housing value, unsurprisingly, were related to the use of lawn chemical inputs. More counter-intuitively, consumers who used chemicals were more likely to believe that such chemicals were bad for water quality, and more concerned about the environment. Pursuing the latter, we talked to people at great length about why they would use such chemicals. Certainly neighbourhood pressure played a role (Robbins et al. 2001). And the normative aesthetic provided by the companies themselves through saturation advertising spurred consumption (Robbins and Sharp 2003). Even so, the most common answer people provided for participating in a multi-billion dollar industry about which most felt either ambivalent or even hostile, is that their 'grass needed it'.

This answer at first seemed appallingly banal. Why do people water the grass? Because the grass tells them to! But as we began to explore this answer, we began to learn a great many things, most of which were hidden. First, it helped us understand the experiential hegemony of the lawn. Advertising and community standards, however forceful, had less of an effect on people's daily feelings, and their drive to act, than did the colour of the grass and the spotty influx of clover and dandelion.

More than this, it helped to explain something about the larger economy as well. For example, rather than continue to obsess about how the chemical economy produced the desire and demand for turfgrass monoculture, we were now able to ask what constraints turfgrass placed on the industry. The answers, requiring a careful analysis of the seasonality of debt payment flows and receipts, of corporate research and development strategy, and of global marketing imperatives, showed that the rate and timing of grass growth, its specific demands in wet and dry seasons, and its vulnerability to inter-annual variability, determined many of the problems for the industry (Robbins and Sharp 2003).

To conceive the question this way, and to research it meaningfully, are two different things. As the project proceeded, we found ourselves rewriting our questions and searching for new forms of evidence. For example, rather than simply query people's attitudes, we began to observe their behaviours, counting and recording the timing and spacing of their lawn activities. These we calibrated against the natural growing cycles of the species in their yards, the vertical and horizontal seasons of perennial grass growth, its dormancy periods, and its reproduction. In the process, rather than the domestication of the grasses in service of the homeowner, we began to see the domestication of the homeowner in service of the grass. The lawn and the chemical economy began to take on a new apparent relationship to the homeowner. Rather than either a 'dupe' of capital whose aesthetics are controlled by industry or a free agent who simply 'happens' to do a lot of lawn work, people began to appear as emergent *subjects* with identities mediated by turfgrass itself.

This view draws on the notion of the subject first outlined by Louis Althusser (Althusser 1971: 182). In 'Ideology and Ideological State Apparatuses', ideology, that taken-for-granted system of belief and practices that make up the economy and society, exists by constituting individuals as 'subjects'. Such subjects, he asserts, are simultaneously actors or agents who act freely – as in the subject of a sentence – while also being subjects of power, who submit to a higher authority. Subjects are both free and fixed, in this sense.

Equally importantly, such subjects need to be 'trained' and 'have their roles assigned' to them in the society and the economy; they need to recognize themselves. This is enacted through the process of interpellation, where the subject is 'hailed', literally named, recognized, and most importantly self-recognized. Most commonly, we think of such interpellation occurring through the actions of social institutions like the church or police, but as Donna Haraway (1997, 2003) has argued, objects and object systems can have this effect too, and diverse people can be interpellated into monolithic systems through the propagation of specific non-humans, 'joint lives' of things and people, with mutually constituted identities.

This is how the turfgrass and homeowner act on one another. Who hails the lawn owner so that they mow, clip, and water? Whose voice do they hear in the morning, when looking out at the grass, they determine that it's time to apply organophosphates or fertilizer? The answer is the lawn itself. The lawn and the industry, in this sense, are in a constant conversation, but one that passes through the individual, creating their complex identity. In other words, not only does an economic understanding of the lawn productively begin with turfgrass as a driver (or quasi-object) entwined in a networked system with its own momentum (or hybrid agency), but also as a participant in constituting who we are, subjectively, and experientially, through interpellation.

Conclusion: doing the geography of human and non-human objects

All of this suggests some straightforward lessons about practising economic geography. First, objects cannot be treated merely as a series of interpretations. While it is indeed useful to perform 'discourse' analysis that reveals the divergent interpretations held by different people about non-humans (trees, disease, lawns), this cannot be the end of social science. Rather, objects are involved in autonomous ways, in the social and economic worlds of people. On the other hand, objects cannot be reduced merely to the attributes, patterns, and effectivities explained solely through research in physical science. Non-humans have social lives, act on economic systems, and have political forces exerted upon them, and are reconstituted in the process. Neither, however, can understanding the object's role in the economy – as a 'commodity', for example, or as an object of 'exchange value' – sufficiently help us to explain the social, cultural, and political life of things. Rather, beginning explanation and exploration with objects as social/physical actors and part of human/non-human networks better helps us to understand the geography of political economic relationships.

There are perhaps two reasons that such an approach to research remains somewhat counter-intuitive to critical scholarship in economic geography. First, it challenges the sometimes deeply held conviction, inherited from particular readings of Marx, that to begin with the object rather than the human relations in which the object is situated is to reproduce rather than challenge the fetish of the commodity. For critical scholars, there is often the supposition that in understanding capitalism the 'social relation between men' (the economy) assumes a 'fantastic form [as] a relation between things' (Marx 1976: 72). This mystification is understood to allow all kinds of pernicious social and ecological effects and so must be avoided. But as the research experiences described here show, the very distinction between 'men' and 'things' required in such an approach, and the concomitant obsession with the discreetness of human and non-human agency, are themselves forms of mystification that are pernicious and might well be brought into question.

The second and perhaps more powerful barrier to research in this vein is methodological. It is difficult to know how to proceed in research and hard to know what an explanation of such relationships might entail or how it might be evaluated. This latter challenge, the methodological one, is by no means the easier because it is less profoundly philosophical. How do we evaluate the agency of players that do not succumb to interviews? How do we read the history of things that do not write? How do we find in ourselves the reachings ('prehensions' or 'graspings', following Haraway 2003: 6) of other things, including physical things like prions or aesthetic things like grassy monocultures, that underline the porousness of our bodies and ideas?

The challenge that such research presents is of course formidable. But the experiences I have related here are intended to show that traces of such objects, networks, and prehensions are all around us, pervading political economies both near and far. It is perhaps simply a methodological issue of deciding where to begin. With our ear cocked in the right direction, it would seem, we should surely be able to hear the sound of nature talking back.

4 SEXING THE ECONOMY, THEORIZING BODIES

Linda McDowell

Twenty years ago, when *Politics and Method* (Massey and Meegan 1985b) demonstrated that theory and method are political, feminist scholarship was virtually invisible in geography. At the time, feminism was a political matter focused on pressing practical issues – the right to control one's body, access to contraception and abortion, male violence against women, the right to enjoy or ban pornographic images, wages for housework, equal pay. In retrospect, it's clear that these demands are academic/economic questions too – about gender segregation, pay inequality, domestic labour, the very definition of the economy itself. And so, feminist scholars began to debate the nature of the economy and the place of unwaged labour within it.

Almost ten years before this I became an academic, fresh from a unique period in women's education. Gender segregation for many was almost complete from the age of 11 to 18 or 21, but girls from families without substantial financial resources were able to benefit from a type of academic education previously only open to those able to pay. In 1960 as the British economy began to boom and the public in general 'never had it so good', I entered a small girls' grammar school, followed by a women's college in Cambridge in 1968. I was taught by women from an earlier generation: never married, independent, absolutely committed to the values of an old-fashioned liberal education. Furthermore, as one of three sisters, the dominant ideology of male superiority barely impinged on my consciousness at home. If you were prepared to work hard, then, it seemed nothing was impossible. Day-to-day discrimination against women barely touched my life, but then I wasn't poor, pregnant or a low-paid worker and the cocoon of the college blunted the impact of a geography department where the tenured academics were all men and a university in which women undergraduates were outnumbered eight to one.

What then led me to socialist feminist scholarship? It is hard to give a definitive answer but a combination of political events, travel, left-wing politics, and experiences in a number of universities (most noticeably the Open University, where the political purpose of academic work was written into its foundations). The influence of colleagues, good friendships, especially through the early years of the Women and Geography Group in the Institute of British Geographers (IBG), becoming a mother, involvement as an academic with events and programmes to increase participation by women and ethnic minorities at all levels in the university, promotion and the responsibilities faced by senior women in institutions still dominated by male decision-makers: all these events have been significant. At the same time, I have been fortunate in having an academic life that paralleled the flowering of feminist scholarship in the academy. As theoretical perspectives shifted, I was part of the development of a relational understanding of the construction of femininities and masculinities, involved in thinking through the theoretical and methodological implications of the personal being political, challenging the very conceptual basis of the social sciences. As part of this, we developed innovative approaches to social and economic questions and emphasized that academic work can change our understanding, if not the material existence, of the structures of gendered patterns of inequality. During all of this time, being a feminist academic has been both a pleasure and a privilege.

It has also been tough, especially initially, when feminist work was derided as 'political' and 'biased' (at a time when objectivity was the standard of good social science). I became used to criticism from referees who saw feminist theories and methods as an anathema, to sexism and hostility from audiences, to colleagues who thought married women had no need of promotion as their husband should be the main breadwinner, to being the only woman on all-male committees. Nevertheless, the personal changes, the support of colleagues and friends in departments in which I have worked, the great interest among many students in feminist work and the now seismic shift in the social sciences as the discursive construction of knowledge is acknowledged have been a delight.

Here I use my own research to map shifts in feminist approaches within economic geography. As second-wave feminist scholars have aged, the focus of their research has changed, from questions about work and careers to issues around childbirth, childrearing, education and health and, more recently, questions about ageing. I too defined my academic focus in part in relation to personal interests: work on professional careers, work on low-skilled, disruptive young men when my son was a teenager, now research with elderly women, much the same age as my mother.

Embodiment matters

Real women do live physical difference in the flesh. (Wajcman 2004: 96)

Wajcman captures a significant association between gender and economic processes, between gender and labour market participation, as well as reflecting the embodiment of social research practices. Men and male bodies are still regarded as the norm in the workplace – whether the rational cerebral masculinity of the high-tech world and the elite professions or the muscular embodied masculine strength of the manufacturing worker. Femininity and women's bodies, on the other hand, are still seen as out of place in many arenas. If women reach the commanding heights – in the professions, the City or running corporations – their age and their appearance continue to be seen as newsworthy. In the years since my book (McDowell 1997) about gendered practices in the City of London, I see few improvements. Clara Furse might have become the first women to head the London Stock Exchange but, as she has noted, 'City news (with some important exceptions) seems to focus on the odd corporate failure, job cuts and, in the case of the London Stock Exchange, merger speculation, the colour of my suits and the cut of my hair' (in Treanor 2003: 26). And a growing number of City discrimination claims show she is not alone in this unwanted attention.

A similar set of assumptions about gendered differences and women's place has been apparent in economics and economic geography. Theories and practices continue to ignore the consequences of these assumptions and of gendered embodiment. Formal theorizing, especially the elegant mathematical modelling of economics and the 'new economic geography', assumes a disembodied rationality that denies the messy complexity of the real world. In the design of research questions and the adoption of particular methods the gendered social processes, structured patterns of inequalities and everyday economic practices that constitute women as inferior economic agents and actors are too often ignored.

Here, then, I focus on the consequences of this absence and try to answer the question: 'what difference does it make to the methods and practice, and to the theoretical propositions, of economic geography if gender relations and embodiment are placed at the centre of investigation?' This is a question that others have addressed (Bergman 1990; Gardiner 1997; Gibson-Graham 1996; Nelson 1992; Vaughan 1997) and, as I have no space here for details, I take for granted a degree of familiarity or the willingness to read beyond this chapter (see McDowell 2000).

But, first, a brief proviso: gender and feminist scholarship is about men as well as women, and about the relationships between them. Gender is a relational construction – the attributes of femininity are what those of masculinity are not and increasingly men and masculinity are the focus of research. In geography, it is important too to recognize and challenge the spatial corollaries of hegemonic assumptions about gender differences. If a woman's place is assumed to be in the home and neighbourhood, if gender is in part about embodiment and bodily processes, then consequentially it is often fallaciously assumed that geographers working on gender relations must focus on the local and the domestic. Gendered assumptions about rights and just rewards structure economic policy at the largest and the smallest scale. Structural adjustment policies, for example, affect the local life chances of men and women. While the examples in this chapter *are* at the local scale, as this is where I work, such a focus does not mean that wider social processes are ignored. The particularity of place is always constituted through the intersection of social processes at different scales and it is mistaken to assume either that gender relations are local or that locally-based research has spatially-limited implications.

Class matters: why economic geography ignored gender for too long

In a world economy currently growing in large part on the basis of women's incorporation into the labour market, it is perverse – but necessary – to have to assert the significance of gender relations in economic analysis. Despite the transformations of economic restructuring, women, in advanced industrial and in developing economies, as a group, continue to occupy different economic positions from men. In general, across the globe, women remain less likely to be part of the waged labour force than men, to work in segmented industrial and occupational groupings, to be grouped, typically, in lower paid and lower status occupational positions and to undertake the major proportion of unpaid work carried out beyond the boundaries of the formal economy, especially domestic and caring work in the home. Clearly something needs explaining. At one level, women are (almost) universally regarded as less eligible than men for commanding positions in the economy because, in the absence of public or private services, they have a continued responsibility for the majority of unpaid caring labour. Women holding the baby and providing services for waged workers simply have less time for waged work.

Behind this division of responsibilities lies an ideological divide between the private world of the home and the public world of the economy that has its origins in modern social theory and in the development of modern social institutions during industrial-urbanization in Western economies. While the responsibilities of men were to engage in the public world of work in order to provide financially for their dependants, women's duties and responsibilities lay elsewhere – in the care for children and the everyday domestic lives of their menfolk. Women then were primarily associated with the domestic sphere and with the unpaid labours of care and social reproduction (Pateman 1988). While sometimes criticized for its ethnocentric bias (but see Ortner 1996; Rosaldo 1980), this argument is an acceptable summary of the basis of the gender division of labour in advanced industrial societies, trailing a long legacy despite women's growing access to the labour market and the public sphere.

A parallel distinction associated with gender is that between nature and culture. Women are associated with nature and men with 'culture', with all the assumed attributes of embodiment and emotions for women and disembodied cerebral thought and activities of the mind for men. While laying bare the implications of these associations has long been a concern of feminists, this work has only recently become mainstream in economic geography. These attributes of embodied emotions, associated with menstruation, childbirth and lactation, and with women's punier physical forms, their inability to think rationally, their very natures, unfitted them for the competitive world of work. And so production is above all a masculine sphere where distinctions between men are on basis of class and status. Men in most jobs and occupations united against women as unfair competitors in the labour market in the development of Western industrial capitalism (Milkman 1987). Indeed, Rosaldo (1980) has suggested that the extreme versions of the public/private ideological divide in the Victorian era compensated men for the insecurity of capitalism. These separate spheres underwrote male authority and risk-taking in the labour market as women's selfless love and devotion confined them to the home.

As feminist scholarship developed, attention turned to theorizing gender in a way that left purchase for challenges to these gendered associations and gave scope to effect social change. First, gender was defined as a construct (rather than a natural phenomenon) in order to capture how behaviours, social practices, rules and regulations and symbolic representations distinguish the sexes (women and the social attributes associated with femininity are constructed as inferior to men and the social attributes of masculinity). This asymmetrical structure is evident in differences in power, authority, status, income and wealth, and social value that accrue to men and women, albeit inter-cut with (socially constructed) differences such as age, ethnicity and sexual preference as well as class position. Since the 1960s feminist scholars have challenged the mind/body: culture/nature dualisms by emphasizing sexuality and embodiment, 'the importance of the body as a physical and biological entity, as lived experience and as a centre of agency, a location for speaking and acting on the world' (Low and Lawrence-Zuniga 2003: 2). Thus, the physical body entered analyses.

In economic geography, embodiment has only recently become a focus of theoretical attention, despite David Harvey's (1998) attempt to argue that Marxist geographers had always theorized the body, as labouring is so often the application of bodily strength to transform material objects. In part, many feminists felt that placing the body at the centre of their work seemed to reinforce arguments about women's bodily confinement. As a young academic I was anxious to assert women's rationality, in attempts to ignore what seemed an almost unchallengeable dichotomy that mapped all too easily on to gender differences. Surely the body could be ignored? But as it turned out, it was part of the explanation.

As feminist scholarship became influential in economic geography, material changes in the world economy and in forms of work hastened the acceptance of feminist arguments. Labour markets in almost all economies were increasingly penetrated by women. Just as the old left (Gorz 1982) insisted on the declining salience of class, the global working class expanded and changed its gender. Deindustrialization in the former core economies of the minority world, paralleled by the rise of service industries, and rising industrialization in the majority world drew women in increasing numbers into the wage relationship. In parallel to the Marxist-influenced regulation theory influential in economic geography during the 1980s and 1990s, feminist theorists posited gender regimes, recasting the links between welfare states and households and the obligations and duties of men and women that are leading to the slow decline of the breadwinner model of social responsibility. A new model – the individual worker model – in which all able-bodied adults are expected to work, is now dominant in the USA and the UK. Now, who cares for children and older dependents, by what means and who pays have become a focus of research by geographers and others (Jarvis et al. 2001; McDowell 2005a). It now difficult to claim, as some regulation theorists in geography did, that because gender relations are outside the capitalist economy, they are outside the boundaries of the subject. Gender divisions are structured and restructured within capitalist social relations and by state intervention in the forms of regulation of the economy, and the household is as much a part of capitalism as any other set of social relations. Relations of care, based on love and reciprocity, as well as, sometimes, on coercion, should be part of an analysis of 'the economic'.

New explanations of occupational segregation: body matters

When I began working on occupational segregation, explanations of women's economic marginalization tended to rely on materialist explanations: the nature of the labour in an area, its time demands, the demands of domestic labour, and ideas about human capital. Thus in work on changes in gender divisions of labour in Great Britain, Doreen Massey and I (McDowell and Massey 1984) drew on a materialist analysis. More recent work on gender and occupational segregation has placed embodiment at its centre: looking at the significance of sexed bodies, as well as the socially defined attributes of masculinity and femininity, turning to different types of explanation and new methods.

It is now widely accepted that bodies are more than biological matter. They are constructed and defined by historical social structures and discourses that define gendered particularities and naturalize a person's existence and role in the world. Thus, inscriptions of socio–political and cultural relations on the body, relations that vary across space, define it rather than biology or biological attributes. Notwithstanding the continued strength of biological understandings of bodies, the work about inscription is yielding new insights in workplace analyses. New connections between economic and social and cultural theory have been developed to better comprehend the interrelationship between class, gender, status and skill, for example in the new cultural economy (Amin and Thrift 2004). Bourdieu's (1977) notion of *habitus* – how social status, moral values and class position become embodied in attitudes, emotions, physical ways of standing and gesturing – is, for example, a useful concept in helping to explain why women are less successful in gaining influential positions, drawing attention to what might be regarded as 'non–economic' traits that nevertheless maintain and explain structures of inequality within the labour market.

There is now a large number of excellent studies at different spatial scales, often drawing on case studies of particular labour processes at a range of sites, from the embodied labour of women in export processing factories and the *maquiladoras* (Cravey 1998; Salzinger 2003; Wright 1997) to data processing centres in the Caribbean (Freeman 2000), domestic work in Canada (England and Stiell 1997; Pratt 2004) and in call centres (Larner 2002; Mirchandani 2004). Using a combination of approaches, links between the social construction of skills, jobs and social evaluations of female bodies are mapped on to each other showing how job evaluations and embodied practices in the labour market combine to reinforce women's unequal treatment. My work on banking came at a time when an intellectual shift in interest from economic structures to notions of performance and gendered daily practices was becoming noticeable. When I look at *Capital Culture* (McDowell, 1997) now, I wonder whether the disjunction between the two approaches could have been handled differently. In part one the approach is a traditional larger-scale analysis of the structures of segregation, in part two I had a more cultural perspective. My next book (2003) shifted the focus from women to men, in part a response to my son's insistence that young men are the newly disadvantaged group as girls outperform boys at school and as job opportunities for unskilled and under-qualified young men shrink. There, too, I tried to combine a materialist analysis of economic restructuring and class divisions with a post-structural focus on the discursive construction of identity. Although I can still see a rather unsatisfactory amalgamation of two different perspectives, I remain convinced that it is crucial to combine structuralist and post-structuralist perspectives – the former insisting on the continuing salience of categorical divisions (of gender, class, etc.), the latter on the importance of addressing issues about fluid gendered performances (see McDowell 2004).

Feminists interested in workplace performances have also adopted new research methods. A focus on embodiment and cultural practices has seen a greater reliance on ethnographic work, participant observation, and in-depth case-study interviews, leading to links between anthropology and sociology rather than the abstract mathematical modelling of economics and the new economic geography. Such research demands bodily presence on the part of the researcher and attention to the relations between the observer and observed, and how the embodiment of both affects interactions. Issues traditionally neglected in work on the economy have to become part of the research process, challenging conventional associations between masculine authority and feminine inferiority. New questions about how to gain access to powerful economic actors and agents have to be addressed which were easier to ignore when face to face contacts were avoidable. Thus problems about how to gain access to elites, often so much harder to study than those with little power and influence, have to be addressed as do issues about what your reception might be when you are much younger (or older) than your interviewees, when you are clearly a different social class or a different gender.

All these questions have been important in my career. I recognize the avuncular kindness of a local authority director of housing which, when I was younger I interpreted as condescension; I am still amazed by the openness of a fund manager who told me more than I had asked as he was unable to see past my pregnant body and recognize a still-functioning brain within! More recently I wonder about the empathy I endeavoured to establish with young unskilled male workers in Sheffield, and with a very different group of elderly Latvian women in Bradford, when my own experiences have been so different from both these groups. How much did my embodiment as a middle-aged, middle-class white woman from an elite university affect the interaction? How can I write this into the work? While more attention is now paid in geography to questions of interpretation and the cultural assumptions that may have structured our research interactions, I still find these questions extremely difficult to answer and almost impossible to address in the sort of academic product we typically construct.

Two urban anthropologists, Low and Lawrence-Zungia (2003), suggested one approach that draws on the work of Alessandro Duranti (1992). In an investigation of the interpenetration of words, bodily movements and the occupation of space in Western Samoa, Duranti developed a theory of 'sighting'. He emphasizes the importance of the connections between bodily gestures, language and spaces which take place in meetings through 'an interactional step whereby participants not only gather information about each other and about the setting but also engage in a negotiated process in which they find themselves physically located in the relevant social hierarchies and ready to assume particular institutional roles' (Duranti 1992: 657). This notion of 'sighting' allows us to conceptualize the implicit summing up that takes place when business partners or clients first meet each other, as well as the similar process the researcher engages in. But analyzing the situation remains a challenge for economic geographers who may have a limited understanding of the implicit exchanges, few methodological tools to analyze the interactions and are anyway typically, although not always, outsiders in the process. Here new work about affect and non-representational theory may have some purchase.

Anxieties about speaking for the 'Other'

In my most recent work, I have begun to look at questions about mobility, transnational spaces and diasporic identities within the frame of a historical case study. I found myself drawn to reconsider the 1950s, in part as a consequence of having been brought up by a 'working mother'. I knew from personal experience that the ideology of a submissive domestic femininity in the 1950s was far from hegemonic and that many women continued to undertake waged labour in these years. I decided to approach an understanding of the 1950s through the eyes of women, like my mother, who continued to 'work' after the end of the war. I knew that migrant women were more likely to be in employment than 'native' women, in part through economic need and sometimes as a condition of entry to the UK. I decided to trace and interview women who had been recruited in displaced persons camps in Germany by the British Government as 'volunteer' workers to aid the British reconstruction effort. These women came to the UK between 1946 and 1949 under two fancifully-named schemes – the Baltic Cygnet Scheme and Westward Ho!

By 2000, it was clear how much my work has been influenced by my experience, age and nostalgia for my childhood years, as well as exciting new work on migration, diasporic identities and gender difference, memory, whiteness, and oral history as method. I wanted to explore the disjunctions between the dominant versions of femininity and women's role in 1950s Britain and the narratives of self constructed by these women who had left their homes in Latvia as young women but who had recreated an imagined community in what they saw as exile in the UK. My original aim to explore the connections between home and workplace participation in the 1950s was challenged, however, by the insistence of 'my respondents' – I hate the claiming in this terminology – that what had mattered to them in their lives was their exile in 1944 (they left Latvia as the Soviet front advanced westwards) and the re-establishment of Latvian independence in 1991. The years in-between seemed relatively insignificant to them, with the exception of events such as the birth of children. These women were elderly by the time I met them in 2001 and 2002 and later I came to understand that not only are traumatic events most clearly remembered but that after about the age of 50, memory speeds up and events that took place during people's twenties remain most vivid.

The publication of this work as a book (McDowell 2005b) made me anxious, in part as I was a newcomer in the field, but a larger anxiety loomed. Am I able to, should I, speak on behalf of these women? Have I judged their lives, their views, accurately? What will the Latvians in Britain think about the ways in which I have interpreted their lives? Should I – a white British woman with no connections at all with Latvia – be the one to speak about the events of sixty years ago, a migration that occurred before I was born but whose participants are still alive? The vexed history of Latvian collaboration with the Nazis also troubles me – have I the right to judge these events? This book, of all the ones I have written, troubles me most. But each time from project formulation to completion, I find myself consumed by similar anxieties and the responsibility owed to the subjects who gave their time and words to me. When I wrote the book about white working-class young men I found it hard to reconcile the respect for them that grew during the research with my academic knowledge that their opportunities were limited and their lives, in the main, would be ones of struggle and economic poverty. But, as critical social scientists, I believe that we also have a duty to uncover and spell out the huge inequalities in contemporary capitalist societies as well as the brave lives of some of the victims, only hoping that what we write might make a difference.

A final comment

I have tried to show here how a feminist economic geography insists on a new agenda for both research and practice. The implications for policy are embedded in the text and I am not sure a ritualized bow to policies such as a minimum wage, equality legislation and so on is needed. But a wider political claim is worth repeating and re-emphasizing: that feminist theory has transformed the social sciences, deconstructing those key dichotomies that influenced our discipline and so excluded from consideration within 'the economic' almost all caring labour that reproduces the current and next generation. Feminist economic geography insists on the inclusion of caring and so on a transformed definition of the economic (Cameron and Gibson-Graham 2003). As Castree (2004) has persuasively argued, the idea of what is included within the economic is not fixed but is rather open to contestation. Castree (2004: 204) issued a challenge to economic geographers to 'take seriously their own role in sustaining, altering or eclipsing the various meanings and referents' of the term, a challenge that I want to support. But, sadly, he ignored the work that feminists have been engaged in altering the meaning of the economic, placing other activities within the purview of the term. A good deal of the new work that now enthuses young scholars – on the ethnographies of workplaces; performance; talk; clothes, words and symbols in constructing economic practices; the connections between class and gender in theorizing spatial divisions of labour; 'non-economic' activities in the home and the locality; exchanges based on trust and reciprocity, but also, it is important to remember, often on fear and oppression – found its first place within feminist scholarship. And while my own work has remained largely at the level of the organization or the locality, I have also argued that similar questions about interactions and trust, about gendered assumptions and responsibilities, about production and reproduction and their connections are important at a more extensive spatial scale. All these issues are now (or should be) a key part of the agenda of a newly invigorated economic geography, which in its promiscuous endorsement of multiple questions and multiple methods surely must continue to delight and confound its adherents and detractors alike.

5 PUTTING PLAY TO WORK

Geraldine Pratt and Caleb Johnston

On 19 March 2004 the play, *Practicing Democracy*, was in its third week of performance. It was playing to a capacity crowd at St James' Community Square, a converted church on Vancouver's affluent west side. This was a small space that brought the audience close to the actors, who were selected as much for their lived experience with poverty as their talents as actors. Theresa, cast as a survival sex worker, began the scene in which she approached Emily on the street for much-needed cash. A woman in the audience shouted "Stop!" She walked up to the stage, asking to replace Emily. The scene began again. Approached by Theresa, the woman denied her the money, but asked her to accompany her to a restaurant for a meal. Theresa scoffed. The woman persisted. Her desire was to help Theresa make some long-term, life-altering plans; to this end, she offered Theresa $20 for thirty minutes of her time. The encounter escalated. Theresa became increasingly and visibly angry, seemingly almost beyond control. She shouted at the woman, told her to stop patronizing her, and to just give her the fucking money. The middle-class woman became upset and began shaking. She felt that she was being attacked. She turned to the director, saying that she wished that she had not come up to participate. The director attempted to calm the situation by saying that Theresa was simply acting the character of Karla, and encouraged the woman by speaking of the powerful scene that she helped to create. He asked the middle-class woman, 'Do you mean to be disrespectful?' Of course she did not. He turned to Theresa and asked, 'Did you feel disrespected?' She did. This moment, which the director judged to be 'one of the most powerful' in all of the performances, created the opportunity 'to get to the real conversation ... which is: how do we bridge this gap which has been created, not just in Vancouver, between the rich and poor. This woman, who wants to help, legitimately wants to help ... has no idea what it is to be Karla [Theresa]' (interview with director, 1 June 2004). At this moment in the forum, the play was momentarily stopped, and the floor was opened to the audience to have this conversation.

We were witnessing an adaptation of Augusto Boal's (1998) legislative forum theatre by Vancouver's Headlines Theatre. The goal of this production was to create a space to devise and discuss creative responses to a now-familiar aspect of neo-liberal economic policy: cuts to social services and welfare provisions. The play emerged from a process that began with a week-long workshop in January 2004, involving 31 participants who applied to Headlines Theatre by submitting a statement about how the welfare cuts had affected them personally.[1] Six actors were selected from this group, and a short twenty-minute play was constructed from stories and experiences gathered during the workshop. The play knit a series of conflicts into a narrative. Once performed, it was immediately re-performed for the same audience, with the expectation that audience members would stop the play when they saw an opportunity to make a positive intervention that might lead to a different outcome. Each intervention was improvised for a few minutes, and the forum was then opened to the audience for discussion.

Our interest in this play turns on the understanding that the stories we tell as academics have effects, and from a desire to tell different stories in different ways to have different kinds of effects. To this end, we are interested in theatre as a means of doing research. It can be seen as a methodological response to recent criticisms that scholarly research over-emphasizes language and ignores aspects of our embodied experiences that cannot be fully represented in speech or writing (Thrift 1997). This is hardly news to theatre artists, who have devised a rich range of methods of working with the body, sometimes as a means of or method for bringing experiences into words.

And because it is ludic and staged, theatre has also been seen as an especially effective and safe space in which to *tell* difficult stories (Houston and Pulido 2002; Jackson 2004; Nagar 2002; Pratt and Kirby 2003). *Practicing Democracy* is an especially interesting case because it brought people who live very different social positions into (sometimes tense) relationships. The intensity of these encounters seemed possible precisely because it was a theatrical production. In the exchange between Theresa and the middle-class women described above, for instance, the director actively protected and fostered the safety of this space by insisting (rightly or not) that Theresa was acting.

Practicing Democracy also can be seen as a remarkable effort to put into practice a participatory action approach. In British Columbia (BC) since 2001 the need has been especially pressing. Under the Liberals, there have been massive cuts to public sector jobs, social assistance, and public services; the minimum wage has been substantially reduced; workfare requirements on young mothers have been stepped up, and childcare supports have been pared back. In March 2003, the United Nations condemned the BC Government for the impact of state policy on women's lives. The speed, scale and scope of the changes were extraordinary, possibly because opposition within government had effectively collapsed. In the face of this, many academics felt the urgency of putting their research to work to educate and generate an informed public debate. As the director of Headlines Theatre put it in his final report, 'Why is *Practicing Democracy* important? Because democratic principles are collapsing all around us' (Diamond 2004: 6).

The play created by Headlines Theatre practised democracy in the first instance by soliciting input into the focus of the play. With the input of two Vancouver city councillors, Headlines constructed a shortlist of topics and, using community networks, community centres and libraries, and media, polled 144 individuals for their priorities. The impact of welfare cuts emerged as the top choice. The director then attempted to work with workshop participants in an open-ended and democratic way to evoke accounts of their experiences of the welfare cuts, and to construct a play that would generate a broader public discussion. The play was performed live to an audience of 1,300 people across all of the performances, and 5,000 viewers saw a televised version of the production. Headlines Theatre's commitment to turning the theatre process into social policy is boldly stated on the cover of the play's programme: 'Using theatre to make law'. This is followed by a letter from the Mayor of Vancouver, pledging that through the project 'City Hall will be able to listen to people who might not otherwise engage in a political process, and incorporate their creative ideas into the development of municipal policy'. We began our study of *Practicing Democracy* because we understood it to be an inspirational case of inserting the raw materials of oral testimony, which many of us generate as researchers, into an elaborated process of theatrical production, public debate and policy change.

But *Practicing Democracy* also raises many practical and ethical issues about the possibilities of, and for, participatory activist research. It is not that *Practicing Democracy* was itself especially problematic; it is its very success that usefully brings into view ethical dilemmas shared with many qualitative researchers.

Mimicking the methodology of legislative forum theatre, we narrate our story through a series of conflicts: one set of conflicts emerged at the workshop in which Headlines Theatre did its 'research' on poverty; a second is tied to our efforts to research the legislative theatre process. Occasionally we shout 'Stop!' and emerge from the narrative to orchestrate a conversation among the various participants in conflict.

Embodied knowledges, veiled disclosures and poverty pimps

Today is the first day that the company has gathered at the Japanese Hall in Vancouver's Downtown Eastside to workshop *Practicing Democracy*. The company of 31 is diverse, with a wide range of ages, ethnicities and social backgrounds. The director explains that the week will be spent using a variety of theatrical techniques to elicit stories and experiences, to 'till the soil' for the creation of a play that will tour several venues in the city as forum theatre. Our shared labour will be organized to respond to one guiding question: how cuts to welfare and social services are creating danger in our day-to-day lives. In doing so, we will be asked to draw upon our own embodied knowledges of poverty.

In the afternoon, the director leads us into our first experiment with image theatre. The company divides into smaller working groups, in which each individual has an opportunity to mould other participants into an image that reflects a moment of their own lived experience. I am shaped into a child on its knees.[2] Above me towers Emily with her fists raised ready to strike, while several others are situated around us with their backs turned in silence. The director then moves around the image, asking each of us to reveal our character's inner monologue in one sentence that begins with 'I want'. When tapped on the shoulder, I blurt out 'I want to know why she hates me', to which Emily replies, 'I just want to provide enough'.

Emily then yells that all she really wants is for someone to listen, to give her a break. She finally breaks down into tears. I am more than a little startled by this sudden outburst, unprepared for its intensity. In the room, there are several moments of slightly uncomfortable silence while Emily collects herself, after which the company discusses the dynamic image and how it provides a site for thinking through the 'oppressed oppressor'. Our group then re-assembles the frozen image, which is photographed with an old Polaroid, posted on the wall for future reference. The director reminds us that these frozen images will form the backbone of our work, establishing the conflict-laden scenarios for the performance.

During the afternoon break on the fourth day, we congregate, casually discussing the week's events, when Michelle raises a concern that there are voices in the group that have been silenced throughout the creative process. By focusing on the more sensational issues of public violence, drug addiction and prostitution, she argues that the company's work has failed to make space for the more 'mundane' or everyday experiences of poverty. Who owns the production? While acknowledging that we have all been hired as paid employees, she contends that Headlines are acting like 'poverty pimps', capitalizing on the experiences of the poor without sharing ownership of the creative process.

After the break, the group gathers together in a large circle in the centre of the auditorium. As the director outlines the afternoon's itinerary, Michelle intervenes, making known her distress about the silencing of stories, issues and experiences. (She tactfully leaves out her accusations of poverty pimping.) The director questions this interpretation. He uses

Maricel (one of the participants) to illustrate the point: while she has been comparatively quiet and reserved, this does not mean that her story is not being heard. If this is the case, she has a responsibility to engage more fully with the workshop. Michelle responds that it is a little ironic that he has used Maricel to exemplify this point, as she is one member who feels particularly marginalized by the process. Maricel says nothing.

At this point, various members of the company enter into the fray. Some are frustrated by the distraction. Some call for more control over the creative process. Others explain the necessity of the director as a central authority. One participant storms out and slams a door in frustration. The afternoon's schedule has been derailed. The director withdraws from his position as director, leaving the resolution of the impasse in the hands of the company. The next four hours are spent trying to resolve the situation.

There was much of methodological interest in the workshop. Theatre techniques were used to elicit narratives – using bodies to evoke stories in ways that allowed economically marginal, and in some cases socially and psychologically vulnerable, individuals to maintain their privacy. Emily's statement, 'I just want to provide enough', came from her experience, but there was no requirement that she reveal details of that experience. In social science research, and especially qualitative research, we are accustomed to requiring respondents to speak in detail about their lives. Tracing the genealogy and scientization of this type of confessional discourse, Foucault (1978) argued that its effects are rarely neutral, in part because the subject comes to understand him or herself through this process of narration.

Trust was rebuilt among participants after the workshop broke down on day four by a round of self-disclosure. But this was a layered and halting process of revealing personal details, and it was not the only means of collecting information about experiences of poverty. Especially among vulnerable individuals, there seems good reason to experiment with different, including indirect, modes of testimony. Significantly, the authority of the theatrical vignettes that crystallized, rather than reported, experiences was never questioned by any of the city councillors or planners with whom we spoke.

Nonetheless, conflict did erupt in the workshop and some participants characterized the process as sensationalizing, undemocratic and exploitative. These are familiar criticisms of qualitative research, especially of projects investigating the lives of economically and socially marginalized people. Concerns about the dynamics of silencing particular people (often women) are ever present in group interactions (Pratt 2001). The pleasures and benefits of witnessing stories of victimization are of general concern. bell hooks (1990: 152), for instance, writes of the victim status demanded of racialized women by white scholars who say 'tell me your story. Only do not speak in a voice of resistance. Only speak from that space in the margin that is a sign of deprivation, a wound, an unfulfilled longing. Only speak your pain.' And, as Emily (a middle-class unemployed welfare worker both in life and the play) said of the play's media reception: 'Nobody's interested in the middle class [who are affected by the cuts]. ... People wanted the fucking dumpster!' [Referring to the play's portrayal of a homeless man living in a dumpster] (Interview, 18 April 2004). We might equally ask, are economic geographers susceptible to selecting the most sensational cases? Are we immune from the desire to tell the exciting story? Do middle-class academics take pleasure in stories of extreme deprivation, in part because such stories simultaneously confirm their own safety and social superiority? These questions seem to be especially pressing within qualitative analyses, in which there seem few controls on extracting the 'compelling' quote or the more provocative storyline.

Two aspects of the *Practicing Democracy* process lent themselves to sensationalization. As one workshop participant put it:

> Believe it or not, those [sensational stories] are the safe images. ... Working with teens in school ... the first improvs you always get are the pimps, the drugs, the alcohol, no matter where the kids come from. ... So those are the safe issues. Those are society's ills. ... [In the workshop, after the breakdown on the fourth day] it became the real consequences. ... It became around kids ... the issues suddenly became real people, as to what you would do, and real situations, and real emotional investment. ... It's much more exciting to be the crack addict on the corner who can't get his fix, so he's going to kick somebody's head in. ... That's emotionally safer to deal with ... because it's more like play-acting. It's more like TV. It's more like Tarantino, for god's sake. (Interview with Louise, 8 February 2004)

The more dramatic story may be the safer story, and respondents may narrate their own story through established scripts about 'society's ills' or other popular culture narratives. The understanding that all of our stories are scripted within narrative conventions has relevance beyond forum theatre. This is the foundation for discourse analysis, but there is a need to be self-reflexive about how discourses structure our own data collection.

The Downtown Eastside location may itself have helped solicit sensational stories; this was something that several participants commented upon. During the course of the week-long workshop, a number of the participants jointly witnessed two incidents of extreme physical violence, just outside or close to the workshop: a fatal stabbing, and a police assault of a young woman. It is not particularly surprising that versions of these events made their way into the final play. No doubt the Downtown Eastside is an extreme context. But these observations about its effect raise the wider methodological issue that what we are told is situational (Pratt 2001; Visweswaran 1994).

Charges of exploitation went beyond the issue of sensationalized representation, and were never fully laid to rest. These are important issues to confront, even if, and possibly especially because, they are unlikely to be easily resolved. Emily voiced very strong views about Headlines Theatre's continuing responsibility toward participants.[3] She was sceptical about whether the potential benefits of the play warranted some of its negative impacts (in particular the trauma that one individual began re-experiencing during the workshop).

> When we ended that play about poverty, about the government and all that shit, and David's asking people to open their heads and their hearts and their whatever. When we finished, Theresa still didn't have a fucking place to live, and was going back on the street. And nobody, including myself – in some way that is totally embarrassing to admit to you or anybody else – nobody cared. We're all a bunch of bull-shitters. Do you know what I mean? I got a fucking job for six weeks [as an actor], Caleb, and I can only hope and believe that City Hall pays attention to some of that, that the experience gave everyone else something, that Theresa finds somewhere to live and Jorge doesn't spend the next six months fucked up at the bottom of a can of beer. … I feel obligated in this weird way to say, oh yeah, it makes a difference, does this, that. But I just don't know in my heart of hearts. (Interview, 7 April 2004)

Researching *Practicing Democracy*

Our first contact with Headlines Theatre was through our applications to join the workshop. Caleb was invited to participate; Gerry was not. We then began a halting e-mail exchange with Jennifer Girard, the community outreach officer of Headlines Theatre, tentatively explaining our interest in the project. It was just as well that we did not know that:

> **Entry from director s journal, January 26, 2004:** We are also starting to see academics circling around the project, wanting to do studies on the participants. It s kind of scary really. Jennifer is getting two to three e-mails a day from people. We said yes over a week ago to a request from the Canadian Centre for Policy Alternatives (CCPA) who approached us first. We know them and want to support them because they do really good work. They are seeking some people with whom to do a two-year study on welfare time limits. We negotiated a one-page document that we gave to the workshop participants and then they will decide whether or not they contact the CCPA. One of them, though, commented when she saw the paper that it was another study by intellectuals to benefit the middle class and threw it back on the desk. She also made the comment that her community (Vancouver s Downtown Eastside) has been studied to death (Diamond 2004 14).[4]

As conversations with Jennifer developed, it was clear that she had many concerns about us researching workshop participants, but that she was intrigued by the possibility of evaluating the project.

> Do we have any way to measure the effect? What if the recommendations that they implement come from multiple sources? Can we cite the project as playing a key role? What if they don t implement them exactly? What if they cause a great stir in the chambers but not in the agencies? Or vice versa? There is a wide area of evaluating here that I would be very curious to have some study of but how to go about it? At this point, we are going on anecdotal evidence. Which is often great, but also often singularly positive. I want to say thank you for saying in your e-mail that you are aware of the possibility of getting in the way of the process   this is a key limitation. But I am open to having a conversation about a study (E-mail communication from Jennifer, 2 February 2004).

We settled on a tentative plan: we would attempt to evaluate the effects of *Practicing Democracy*. Jennifer had been astute in her musings about the methodological challenges. Our approach was to conduct interviews with some of the participants (whom Caleb had come to know through the workshop) and different 'actors' at City Hall, together with a close analysis of reports written by Headlines Theatre and planners at City Hall.

Headlines staged two major performances at City Hall after *Practicing Democracy* closed: one was to present the report of the recommendations that came out of the forum theatre, another was to respond to the Planning Department's response to the report. It was essential to both performances that there be visible support for the project. An e-mail call from Headlines requested that supporters register (in advance) for time to speak to City Council. Neither Caleb nor I made the call (although we attended the council meeting). Were we too busy, too lazy, or too concerned with our public face of neutrality? It's hard to say. Later, Caleb called and e-mailed the director of *Practicing Democracy* for a further interview. He did not return our calls. Has he been too busy, too lazy, or did he feel that we had let him down? It's hard to say.[5]

We were drawn to *Practicing Democracy* as an ambitious attempt to use the raw materials of a kind of poverty research to make immediate policy change. We wanted to participate in, support and study the process at close hand, in part to provoke ourselves to think more creatively and learn about a broader range of methodologies and research outcomes. But, as our halting relationship with Headlines makes clear, working as a researcher–ally is different from doing activist work. There are some good reasons for understanding and maintaining this distinction.

As Jennifer noted, the evaluations that come back to Headlines Theatre are anecdotal and 'often singularly positive'. As researchers from a university, workshop participants, actors, city councillors and social planners were likely more willing to express criticisms of the process to us than to Headlines Theatre. The first question posed by one of the city councillors who was seen (rightly or wrongly) to be among the least supportive of the project was: 'Are you involved with *Practicing Democracy*, or you just saw it as an interesting project?' While honest in our response, we made every effort to establish sufficient distance to allow him to voice his criticisms.

Gerry:	A bit of both. I guess we saw it as an interesting project and Caleb then got involved during the workshop as a way of being part of the process, from the beginning – as a kind of ethnographic approach to seeing the process through...
Caleb:	Yeah, the workshop creation and development.
Gerry:	But we're also that much more distanced studying it. I've done other work looking at popular theatre in Vancouver. Caleb is looking at the use of this legislative forum theatre methodology in India [as well]. That's one way in which the geography comes into it – the way in which the same cultural methodologies or forms get taken up and used in different ways in different places ...
Councillor:	How am I going to help you?

This is not to say that we were neutral, detached, un-situated observers. But we could gather a range of opinions and stories, including some that were critical of the process, because we were not perceived to be totally invested in its success. This is important to the process of evaluation.

Whether *Practicing Democracy* can be considered a success requires a more nuanced and sustained analysis than is possible here. Following the *Practicing Democracy* report, the City Council took two actions. One was to disperse a number of the recommendations to four existing task forces or committees. A second was to explore the feasibility of creating a Homeless and Sex Trade Worker Advocate's position. The results of these initiatives are unknown at this date.

None of those we interviewed at City Hall questioned the veracity of the experiences portrayed in the play. Rather, they spoke of the difficulty of translating forum theatre (which all judged to be a success) into social policy. They found some of the recommendations unrealistic in terms of local government mandate, existing private property law, or a wider set of implications. One councillor doubted that a representative sample of the population had attended, questioning whether the recommendations expressed the wishes of the majority of Vancouver residents. Some councillors spoke of the project's limitation as a 'one-off' event, and of the need to deepen the process of citizen input by fostering a more sustained culture of participatory democracy:

> [*Practicing Democracy*] challenges people in ways that may not have been anticipated, but they walk out of there having had a discussion and there is no re-entering that dialogue. ... In terms of community, it doesn't have that same regenerative discussion that, within the situation of the *favelas* in Brazil, was part of Boal's and Freire's dynamic. The literacy process kept them at the table and seeking solutions constantly. (Interview with Councillor David Cadman, July 2004)

As researchers, we see ourselves entangled and in solidarity with activists, in this case popular theatre activists, in a process of generating and fostering this lively public culture of debate and participation. We are able to contribute to this process precisely because we are aligned with, and not identical to, the activists with whom we are in solidarity. We have brought something else to the debate.

Research as play

In many senses, we simply tagged alongside an intriguing experiment in legislative forum theatre, to glean ideas about new techniques to elicit and tell stories about living in poverty, and to open our imagination to what activist research might aspire to achieve. Not surprisingly, familiar debates about exploitation and domination surfaced within this project. We have left ourselves and the reader in the middle and the muddle of these debates, because we believe that this is where we should remain. Every research project must of necessity muddle anew through the same old debates. But we also brought to this alliance our skills in discourse analysis, our privileged access to a diverse range of opinion, and our significant resource of time to research and write.

As researchers, we can gather stories, contextualize them, analyze the narrative structures, societal scripts and geographical sites within which they are crafted, and deliver them across points of conflict to generate a deeper public debate. Our 'distance' as academic researchers opens some non-trivial parallels to the safe and ludic space of the theatre, as a role and site where different and sometimes conflicting perspectives come into contact and are put into conversation. This is our potential as researchers to practice democracy. There are few limits on how we might choose to develop this potential: *Practicing Democracy* offers but one model to excite our imagination.

NOTES

1 Caleb Johnston participated in this week-long workshop. Subsequently, we have interviewed seven of the workshop participants, two members of Headlines Theatre, the 'legislator', three members of city council and one social planner.

2 The 'I' here refers to Caleb.

3 Concerns about exploitation must be balanced against the resources and support that Headlines Theatre assembled for participants. All workshop participants were paid an honorarium of $500, while cast members received $550/weekly over a seven-week period. A support worker was on staff. Headlines also provided bus tickets, a childcare subsidy for single parents, and other supports. Safe accommodation was arranged for several homeless participants. The allocation of the production budget of approximately $120,000 clearly articulated Headlines' awareness of the precarious living circumstances of many (not all) cast members and workshop participants.

4 Yet Caleb did not sense a general antipathy to outside researchers among the workshop participants. There may be a danger of over-generalizing the comments of one participant.

5 Communication with the director has since been re-established and he has been generous with his time in discussing our draft analyses. Appropriately enough, communication was re-established when we had something concrete to offer.

6 OF PUFFERFISH AND ETHNOGRAPHY: PLUMBING NEW DEPTHS IN ECONOMIC GEOGRAPHY

Elizabeth C. Dunn

Economic geographers are increasingly concerned with the ways in which decisions made in smaller places – in regions, industrial districts, and even inside workplaces – shape larger economic trends. But do we really have the tools to connect the dots between the choices and actions of individuals at the local level and larger-scale economic formations? I would say not. The problem, I think, results directly from a misunderstanding of the terms 'culture' and 'ethnography', a theoretical lack of attention to both culture and economy as lived processes, and a resulting failure to develop methods to observe those processes.

Put more cogently, the problem with 'qualitative research' in economic geography is that it is mainly defined as 'doing interviews'. Interview-based research certainly has strengths: talking with economic decision-makers is an important way to gain insight into the strategies of elites (Schoenberger 1991), and interviewing people victimized by large-scale economic processes gives us an acute picture of their plight. But interviewing is inherently problematic, because the stories people tell about how they make decisions are often radically different from the ways those decisions were actually made. What qualitative economic geography needs, I think, is a wider array of methods and techniques that allow us to see what people do, as well as what they say they do.

My own journey as an anthropologist-turned-geographer offers a few ideas about alternative research methods. As part of my own research, I have driven a forklift, developed an employee evaluation matrix, operated a bottle-capping machine, helped site a silicon chip factory, waded ankle-deep in wet cow manure, and slaughtered a pig. This kind of participant-observation reveals the multiplex forms of informal action that shape economic behaviour, the recognition of which can be combined with more traditional economic-geographic techniques (like sector analysis or formal interviewing) to produce better explanations. This, in turn, has very definite political implications: by revealing the day-to-day practices of all kinds of workers, we can expose the political stakes of economic practice and open up arenas for political contestation.

Pufferfish, or, why asking isn't the same as doing it yourself

Consider for a moment the pufferfish. Pufferfish are smallish fish, sleek and tube-shaped, and so slow they can often be caught by hand. They are quite tasty, which makes them juicy prey for larger fish seeking a floating hors d'oeuvre. When the humble pufferfish is threatened, though, it has an extraordinary defence: It quickly gulps a huge quantity of water and blows itself up to four or five times its previous size. Its body becomes almost globular, making it difficult for predators to grip the fish long enough to kill it. Hard spines, previously concealed against the sleek body, spring out to deter bites or to wedge the fish in a hidey-hole. The once-delicate meal becomes decidedly un-fish-like in the eyes of predators, who move on in search of easier prey.

Many of the people we interview – particularly managers, a favourite target of economic geographers – are pufferfish. Seeing researchers as potential threats, they may inflate themselves to seem more intimidating. Particularly when discussing a controversial topic, they practise their stories until they are hard, polished nuggets of rehearsed text, leaving the researcher little to do but write down the pre-digested narrative as it is told. Consciously or not, the people we interview often provide *ex post facto* justifications for decisions that weren't really made that way at all, rearrange facts to make the story better, leave out conflicting opinions or attempts to change the course of events, disregard crossroads at which other outcomes might have been generated and generally leave the impression that the decision that *was* made was the only decision that could have *possibly* been made. But this is often a misrepresentation of the actual processes of deciding, building, and creating. In an interview, even friendly interviewees inflate themselves and craft stories that act as spines, making it extremely difficult for the researcher to get any grip at all on the actual processes that drive that part of the economy.

The implicit idea behind interviews is that the subject's narrative can stand as a proxy for reality. Such 'texts' become flattened and iconic representations of lived reality, something that can be carried off to the confines of one's office and pored over for hidden meanings, the assumption being that texts provide windows on reality. Sometimes those windows are taken to be totally transparent, and the very fact that what we are studying is a representation is completely glossed over. Other times, in more critical works, the 'distortion' of representation is acknowledged, and the window becomes partially hazy. But even acknowledging the distortion does not solve the fundamental problem of the second-hand information that comes out of interviews: we are limited by our informants' understandings. This is not to denigrate interviews as a method. Interviews with pufferfish can tell us a great deal about people's ideologies and their visions of the world they inhabit. They can tell a lot about ideal-typical models – about how things *should* work. But formal interviews often tell us very little about how things really *do* work.

Thrift references this problem when he talks about the ways that geography has focused on the event. Such 'representationalist approaches', he argues, tend to be preoccupied by 'life from the point of view of individual agents who generate action' (2000: 216). This flaw is inherent in the very nature of the interview, which sees talk-as-text as a stream of signs that express 'culture'. The interview fails because it is a *representation*, an act of signification, which necessarily obscures practical activities. That tendency is clearly evident in economic geography, which very often places human agents (whether workers or managers) as the heroes of the narrative. Even more macro-scale versions have non-human heroes and villains (such as markets) which are nonetheless endowed with the anthropomorphic capacity for purposeful action.

Using the narrative text as the unit of analysis also leads to a specific concept of time, one in which events follow one another sequentially to a preordained conclusion. But, as Thrift points out, the world is an uncertain place: one which is 'howling into the unknown' because the consequences of actions are not always known in advance and the actual outcome lies in a field of unrealized – but entirely possible – alternatives. Events have a 'surprisingness' because there is always a surplus that makes them something more than the predictable outcome of previous events (Thrift 2000: 217). To grasp that surplus and to understand events as processes, rather than foregone conclusions, we must go beyond the text to seek out embodied practices and the ways that people draw on often-unstated rules and presuppositions. We have to look for the kind of 'culture' that exceeds self-conscious representation. From this perspective, economies appear less as the structural residues of rational action, more outcomes of particular sets of practices that remain inchoate, unspoken, and *sub rosa*. The only way to get at them is through work: not talk about work, but work itself. We have actually to do the job.

The underground (il)logics of work

The study of illicit practices is a beginning foray into the kind of research that might transform 'culturalist' economic geography. 'Illicit' here means something both more and less than illegal. It means anything not expressly authorized by the work system itself: the shortcuts and work-arounds that people sometimes use to foil the production system, but more often use to make the system work in spite of itself. Sit-down interviews can't tell researchers much about circumventions, fudges, negotiations, or compromises, even though these informal practices not only shape actual production, but also corporate strategies, regulation, and the physical landscape.

If economic geography is the study of value in space, neglecting informality means neglecting important sites of value production. At one scale, the neglect of informality can shut us out of entire world regions where informality goes far beyond a mere greasing of the bureaucratic wheels. Böröcz (2000: 352) suggests that in post-socialist Eastern Europe, the 'moral predominance of formality' that characterizes Western European and North American societies is replaced by hyper-endemic informality of all sorts, including graft, 'mafia business', the use of personal connections, underground employment and rule circumvention. In many of these economies, the bulk of all economic activity is illicit, or at least informal. The same may be true, in different places and times, of Latin America or the Middle East, or any other area of the world in which late industrialization and weak states mean that formality and rule-driven governance are not norms of social conduct, as they are in the West.

Ignoring informality may also blind us to places at smaller scales where important economic activity takes place. Take food safety and the regulation of the meat industry, where *formal* activity takes place in public cattle auctions, US Department of Agriculture public notice and comment meetings, and corporate management offices, where food safety plans are developed. However, the industry is profoundly shaped by informal activity taking place in back rooms, on feedlots and on the floor of the slaughterhouse. At one symposium on food safety that I attended, I entered the ladies' room only to discover where the real negotiations were taking place.

Charles Tilly (1997) has a beautiful metaphor to explain the role of informality in the economy: he calls it 'the invisible elbow'. While people in power often give an *ex post facto* account of their actions to make it seem as if they were totally governed by rational-technical logic, what really happens is quite different. Rather than executing a well crafted plan, social actors are forever attempting to do things, making mistakes, and self-correcting their own errors. It is this rough-and-ready kind of action, 'the invisible elbow', that shapes the economy, not the rational 'invisible hand' of self-interest. 'Culture', in this context, is not simply an exogenous constraint; it is the hardened residue of solutions arrived at by trial and error, negotiation, conflict and compromise. Culture is thus forged both in and through formal meetings and processes *and* in the restroom and on the back of the envelope.

If we are to understand the lurching series of decisions, responses, unintended effects and fixes that constitute the economy, we have, first, to take the process of economic action – with all its uncertainties and interactions – as seriously as the outcomes of economic action, or the *post hoc* narratives that actors tell about them (and this requires getting into the actual spaces where 'the invisible elbow' is at work); second, to change the time horizons of our research by suspending, at least momentarily, the drive for a 'genetic' explanation of an event, and strive to understand the present in its own right (we need to see not only what course of action is chosen, but those potential paths that are not chosen, and why); and third, we have to take the context of improvised decisions seriously, since it is both a constraint on structural change and the productive soil in which that change grows.

But how do we dredge the informal, unofficial and even illicit parts of the economy? Like anthropologists, economic geographers need to submerge themselves in their investigations. Anthropologists discovered the utility of participant observation after decades of research done from the comfort of an armchair, or from tents in the field where informants were brought for formal sit-down interviews. After the Malinowskian revolution, anthropological research became premised on three distinct principles. First, the claim to having 'been there' and observed the practices of the locals firsthand; second, a long-term engagement with local communities, rather than the brief encounters interviews provide; and third, a willingness to understand the nature of the work by experiencing the work through the medium of one's own body, that is, by joining the research subjects in putting shoulder to the wheel.

Aihwa Ong (1987), for example, watched her Malaysian co-workers in the semiconductor factory fall to the ground, possessed by spirits. Evidently, the workers were being subjected to rigid Taylorist discipline in order to improve their productive output on the line. Nobody, not even a pregnant woman, was permitted to leave the line to go to the toilet. But women who were possessed by spirits, dropping to the ground and frothing at the mouth were quickly hustled away to a place where they could recover out of sight – the ladies' room. Michael Burawoy (1979) detailed the gymnastics machinists had to go through in order to 'make out' under the piece-rate system and motivate themselves to continue producing at a furious rate. The mental and social games they played ran the gamut from negotiating with the scheduling man to sneaking poor-quality product past the inspector. Likewise, Miklós Haraszti (1977) showed the ways that workers made their labour meaningful by swiping raw materials, tools and time from the factory in order to make 'homers', or things for their own consumption. These events rarely show up in interviews. Few managers, when asked about the origins of their decisions to stop contracting labour in Southeast Asia, would mention spirit attacks as part of the rationale for their decision. Few workers, even when interviewed by sympathetic researchers, would mention making homers. It is only insiders, those who participate in the day-to-day grind, who know the tricks of the trade and their social meanings.

And working alongside workers produces a different kind of learning and knowledge. On the job, researchers obtain a different kind of knowledge: a kind of practical embodied knowledge that is the basis for the quick adjustments on the fly. This kind of knowledge, which Scott (1998) calls *mētis*, is what makes Tilly's invisible elbow a roughly efficient means of closing the door, rather than just a flailing gesture. It is the basis of the art of satisficing, of making do, of generating approximations that are not precise or even based on full knowledge, but that aim to get the job done. Without understanding this rough-and-ready logic, it is impossible to understand how economic events really happen.

Bulletproofing, negotiation, and work-arounds: audit in practice

Managers often use the same sorts of 'informal' techniques in their jobs. The tension between formality and informality in managerial life is mediated by *audit*, a technique that disciplines by requiring disclosure, but which also eclipses or hides entire domains of social action. Power (1997), for example, argued that financial accounting systems drive corporations to produce 'auditable documents' that managers use to constitute the company's activities. Rather than opening a transparent window on to the corporation, accounts are the product of long negotiations by managers, accountants and auditors, all of whom modify and reinterpret formal accounting rules. Strathern (2006) compares the documents that organizations produce to the shields created by tribal peoples in New Guinea where the men paint their own faces on their shields, but as terrible and fierce facsimiles, designed to convince the enemy that the man behind the shield is so frightening that it is better not to attack at all. Managers use the auditable documents they produce in just the same way, filling out documents both to prove to auditors that the organization is compliant and to create an image of the organizational self that will convince the auditors not to look behind the shield of the document.

In packinghouses, for example, the US Department of Agriculture (USDA) now uses a system called Hazards Analysis of Critical Control Points (HACCP) to monitor the disassembly process and to ensure the meat is not contaminated. According to this rule, packers may themselves define potentially hazardous points in the production process, and procedures or remedies to prevent contamination, but they must prove that both the risk assessment and containment measures are 'scientifically valid'. 'Science' here stands as the link between the text of the auditable plan and an objective reality: it signals a correspondence between the world of the document and the world of the carcass so tight there is no room for negotiation or interpretation. Companies use trial and error to determine which kinds of production process meet with the USDA's approval – and that varies from plant to plant, inspector to inspector. In one plant, the owner installed a very expensive steam cabinet to superheat the carcass and kill germs. The Federal Inspector demanded that he rip the cabinet out, though, saying that it had not been 'scientifically verified'. The two men began a long series of verbal and practical negotiations. The packer marshalled different kinds of evidence, reconfigured the cabinet and the line several times, and numerous tests to prove the cabinet's effectiveness, but ultimately concluded that 'nobody knows what the science is'. The problem was finally resolved, not by a definitive scientific answer or a decisive victory on the part of the packer or the inspector, but by compromise: the inspector agreed to allow the cabinet if the packer made technological changes elsewhere on the line.

Auditable records act as bulletproof shields precisely because they efface the negotiations and work-arounds that exist in the production process. For the packers, the documents function as a shield against lawsuits. They are proof that the producers have been 'duly diligent' in their quest to eliminate food-borne pathogens, and hence cannot be held liable for consumers' illnesses or deaths. For the USDA, auditing HACCP plans and production records gives the 'proof' that the agency has gone to a great deal of effort to control the meatpacking industry. This is politically essential, given that the agency has been portrayed as a captive of the industry and has been widely lambasted for failing to protect the American meat supply (see Schlosser 2001). The auditable document gives a smooth, hard shield that the industry can hold up to interviewers ignorant of the vagaries of the actual production process.

Some may scoff at the microscopic attention to detail this kind of research requires. Who cares if the numbers on a production record are jiggled upwards a little, or ratcheted down just a hair? The issues seem arcane, if not irrelevant, and certainly not worth the amount of labour it takes a researcher to explore them. In aggregate, however, these tiny adjustments can have monumental effects at the macro level. Anyone who doubts this has only to look at the political economy of East European state socialism. In the command economy, the architects of the Five Year Plan gave managers of state-owned enterprises a specified amount of inputs and concrete targets for outputs. Managers, however, did not accept the numbers on the plan as givens: they used their relationships with their superiors to 'bargain the plan'. For example, managers often inflated their requests for inputs beyond what they actually needed, because they wanted to be able to stash resources to trade with other firms in the sector, which in turn had hoards of their own. Employees on the shop floor engaged in practical negotiations and work-arounds. They often hoarded materials in order to meet their own production targets or stashed the factory's materials away to make the 'homers' that Haraszti (1997) describes. Because they often did not have the materials to do the job right, workers often had to improvise, taking shortcuts in the production process and using techniques or materials that were not specified in the plan.

The result was not only poor quality product, but also enormous waves of shortage. Plan bargaining, stealing and hoarding caused shortages, which led to further hoarding to cope with future shortages which, in turn, led to more acute shortages. The downwards spiral of shortage and hoarding became politically unsustainable for a regime claiming to be able to meet the basic needs of the entire population. While other factors certainly came into play (the arms race, for example), this basic dynamic made state socialism vulnerable to internal dissent and, eventually, to economic collapse. It was a collapse that few Western researchers predicted. Unable to travel freely in Eastern Europe, and hence unable to understand the true dynamics of production, Western analysts had focused on the central bureaucracy and scoured official proclamations in an attempt to read the tea leaves. This research approach – one step removed from the formal interview – left the true motor of economic change undiscovered and unexplored.

Ethnography and knowledge: cultural externalities

Clearly, what I am espousing is 'doing ethnography' in a rich anthropological sense, one that goes beyond the currently fashionable, bowdlerized version of that term that reduces it to mere talk. But I have also used the term in a restricted sense, confining ethnography to the workplace and the production process. This, I think, leads to an overly simplistic dichotomy between practical and system-formal knowledges, and hence misperceives how ways of knowing in the workplace shape economic action. I want to repair that unfortunate gap by introducing a third term, and a third kind of knowledge: cultural externalities.

We might think of this problem by returning to the problem of textuality. Interviews and auditable documents give us *texts*, planned self-representations. Participant observation gives us a standpoint outside the bureaucratic system that produces the text from which we can reflect on the system itself. By contrasting system-formal knowledge to practical knowledge, we can begin to look at bureaucratic systems (like HACCP, for example) as *paratexts*, in the same way that physical infrastructures like roads, pipes, or software are paratexts. Practical knowledge does not encounter these paratextual structures and remain unchanged. It is profoundly shaped by the drive to fit into, modify, or circumvent the categories of system-formal knowledge. We might think of the resulting hybrid form of knowledge – one that is called forth by the system but not created within its boundaries – as, to coin a term, *isomic*. But this, too, is bounded and assumes that all of the rationalities that guide action are constructed in the workplace, and that other kinds of knowing, doing, and being don't enter into workplace behaviour. The idea of 'corporate culture', much discussed in the 1990s, replicates this intellectual mistake.

The problem, of course, is that managers and workers draw on rich libraries of habits, values, norms, and other patterns of behaviour from outside the firm in their actions within the firm. This kind of knowledge is *metatextual*: it is a context and a resource that people use as they produce texts and as they deploy isomic knowledge to circumvent paratexual infrastructures, but it is not necessarily apparent in either the system or the negotiations that happen in practice. Take, for example, the case of village slaughterhouses in Poland (Dunn 2003). Following Poland's accession to the European Union, small village slaughterers had to adopt the EU's audit-based health and safety protocols and produce texts on a regular basis that ascertain their compliance to those norms. While many of them bribe inspectors or fudge the numbers on their records, others have been forced out of legal markets, trading meat illicitly, without the imprimatur of formal inspection. They use the conventions, spaces, and networks of the suitcase trade, which was built up over decades during the state socialist period, when black and 'grey' markets were rife. These illicit conventions, spaces and networks are all part of the *metatext* of state socialism.

Economic geographers have tried to incorporate metatextual knowledge in various ways. The most promising, but least successful, was the notion of the mode of social regulation, which regulation theorists saw as the social norms, values, and habits that govern economic actions (Aglietta 1987; Boyer 1990). Unfortunately, regulation theory largely focused on the macro scale of the 'regime of accumulation'. The mode of social regulation remained a black box and the French regulation theorists did very little to understand in concrete terms how economies functioned as social formations with their own – often surprising – trajectories. Seeking to remedy this failure, other geographers ditched the jargon and sought to explore the effects of 'culture', *tout court*, on firm behaviour.

Schoenberger (1997: 120) saw culture as 'material practices, social relations, and ways of thinking. Culture both produces these things and is a product of them in a complicated and highly contested historical process.'

While she acknowledged that culture was partial (in so far as there are multiple 'subcultures' within a given firm) and exceeded the boundaries of the firm itself, Schoenberger's working notion of corporate culture saw it as constructed by the dominant actors in a firm (i.e., managers) and, more or less, bounded within the firm. She examined a series of important mishaps in which managers, seeking to defend their own sets of acquired knowledge, power and skills, simply refused to see threats to the firm that did not conform to their views of legitimate competition, even when they were told about the threats point blank.

Xerox, the copier company, was a case in point: the firm lost over 75 per cent of its market to Japanese competitors in the space of ten years, even though its own employees were warning top management of the Japanese threat. For Schoenberger, top management could not perceive the threat because they were focused on 'normal competition'. In Schoenberger's account, the habits, values, and norms that managers rely upon and the assets they seek to defend are part of their identities as *managers* and play out only in the business world. But there were other terrains on which competition is played out, and these may have profoundly shaped business strategy. One was Rochester, New York, the company town that was dom- inated by Xerox and Kodak. The two firms organized not only the copier market, but the social world of the town. Xerox and Kodak executives fought over places in the social hierarchy, from seats on the board of the local philharmonic to country club memberships. They had 'friendly' com- petitive golf tournaments. There were Xerox neighbourhoods and Kodak neighbourhoods, and managers who took up residence in the wrong part of town might find a large 'X' or a 'K' burned into the lawn. The division of the social world and the competition between the two firms, then, went far beyond the copier market: it pervaded social life for managers, their wives, and even their children. No wonder, then, that Xerox executives had their sights so firmly set on Kodak as the 'real' competitive threat. Everything in the spaces Xerox managers lived and worked in pointed to Kodak as 'the other', not the Japanese.

My point is obvious: 'Culture' as it exists outside the workplace can be as much a factor in corporate behaviour as formal systems, practical knowl- edge, or identities as they exist in the workplace, even though that kind of 'culture' is often seen as external to rational economic decision-making. It enters into the matrices of bureaucratic categories, auditable documents, quarterly reports, and stock prices. 'Culture' in its largest sense infuses prac- tical knowledge, as common-sense understandings are built into work- arounds or circumventions. It can be the font of innovation, as the habits or structures from one era are recontextualized in another. It is this rich interaction between events in the moment that they are happening and the cultural and social infrastructures that surround them that give economies their unpredictability, their sense of hurtling into unknown futures, and their deeply local variation. To discover how place *really* matters to economies, we have to incorporate these views of culture and time into economic geography and create 'thick descriptions' of the lived world economic actors inhabit (Geertz 1973).

Ethnography in stormy disciplinary seas

Recently, anthropologist Marilyn Strathern has bemoaned that 'ethnography' has been picked up by many other disciplines, where it is often reduced to 'nothing more than talking with people'. The result is that the 'method simply does not carry weight; precisely to the extent that bowdlerization of ethnography renders it trivial' (Strathern 2004: 554). If anthropologists fear the loss of their key method, and feel that its transfer to other disciplines comes with the sacrifice of explanatory power, why argue that economic geography crosses disciplinary boundaries and verges into anthropology's territory? What, precisely, am I hoping for by tossing the word 'culture' out again, and hoping that economic geographers will engage in some form of ethnography in order to discover it?

What I am calling for is more than a qualitative revolution. It is a revolution in intensity and method within economic geography. In contrast to a 'drive by' ethnography, which features quick trips to 'the field' and hour-long interviews in offices or cafés, economic geography needs to legitimize and even *demand* a longer, deeper, more intense engagement with the subjects of the research. Rather than assuming that they share a 'culture' with European workers or American managers, economic geographers should approach them as if they were a tribe of Amazonian Indians or a monastery of Tibetan bellringers – as a group so profoundly foreign that understanding them requires months or years of immersion into their language, rituals and concerns. What I am proposing is *fieldwork*, in the Malinowskian sense – an attempt to understand an entire lifeworld. This will not lead to the rapid publication of minimum publishable units, but is the only way to get beyond the hard, slick stories informants tell us, abandon the notion that economies are overdetermined, and understand why economic behaviour takes the surprising turns it does. If we are to understand the world of the pufferfish, we are going to have to stop surfing and start diving.

Section 2

Politicizing Method: Activating Economic Geographies

7 METHOD AND POLITICS: AVOIDING DETERMINISM AND EMBRACING NORMATIVITY

Andrew Sayer

I suspect that, like the book which inspired the current collection, *Politics and Method* (Massey and Meegan 1985b), this volume's contributions will be mainly about method, with politics mostly implicit. It is not that the contributors regard politics as less important but, now as then, there is a gulf between politics and social science that is difficult to bridge. The normative character of politics apparently makes it incompatible with the positive business of understanding how the world works In this chapter I will comment on the relation of normativity to social science and argue that the divorce of normative and positive thinking about economic life is unnecessary and damaging. I begin with one of the main debates of the last twenty years about method and metaphysics – on (anti-)essentialism. Anti-essentialists regard essentialism as a narrowly deterministic view, obscuring the potential for intervention and change. Space limitations allow only a limited answer to this; all I can do is sketch out the terrain and reference fuller discussions I begin with (anti-)essentialism and then discuss how ethical–political argument relates to social science.

Method and metaphysics: essences and determinism

The term 'essentialism' is used in many different ways, often sliding between logically-independent meanings. As a result, 'anti-essentialism' is directed against a number of different targets, often producing confusion. I have discussed the debate elsewhere (Sayer 2000a), and so I will make only some points relevant to possibilities for social change.

When we denote something as possessing an essence, we claim (fallibly, of course) that it has certain essential properties – ones that make it that kind of object rather than another. These are distinguished from its accidental properties – which it may or may not have, but which don't make it a different kind of object (O'Neill 1994). Thus the essence of water might be described as H_2O, a molecular structure which gives rise to properties such as freezing, while being in a river or in a lake are accidental properties. It is hard to imagine making much sense of the world without such distinctions.

In addition, to talk of the properties of objects is to refer to their causal powers and susceptibilities, that is, their capacities to produce or suffer change. However, these don't have to be activated. Further, if they are, the results are always context-dependent. In reading this chapter you are exercising your acquired linguistic causal powers, but much of the time you are not doing so. When you do read, moreover, the effects depend on the causal powers and susceptibilities of whatever constitutes the context. Or again, capitalism can operate in many countries, but in each case what actually results is mediated by the context. Further, the very existence of particular objects and processes that might be considered to have essences is dependent on conditions whose presence is neither necessary nor impossible. There is nothing inevitable about their continued existence, and consequently essentialism is not deterministic.

Many fear that attributing essences, as is implied in talking of 'human nature', will lead to a radical underestimation of contingency, novelty, variety and hence the scope for social and political change, affirming the *status quo* rather than being emancipatory. It may do, but not necessarily, for as we have seen, essentialism does not entail determinism. One could equally claim that an essential feature of human beings is that they are susceptible to cultural variation and are reflexive. Hence they are able to intervene in and alter processes, including their own development. In this sense, an essential feature of humans is that they cannot be understood deterministically. Even at a physical level, human nature has a significant degree of variation. For this reason, we need to be open to non-deterministic and inclusive conceptions of human nature sensitive to physical variety, our deeply social nature and our universal capacity for cultural variegation (Dupré 2001).

The obverse of essentialism, anti-essentialism, has been especially dominant in feminism. This is not surprising given that gender has no essence: it lacks a stable, uniform fixed set of characteristics. Rather, the term refers to common bundles of associations and contrasts and axes of domination that are contestable and shift continually across space and time. However, it does not follow that because gender has no essence, nothing has any essence. We should avoid the dogma that either nothing has an essence or that everything does. Even though gender does not have an essence, particular instances of it, such as working-class masculinity at a certain time and place, have particular tendencies, powers or susceptibilities, which produce effects, such as the use of particular cultural practices (e.g., beer drinking) or particular conceptions of women. Wherever there is change there are causal powers producing change. Causal powers include actors' concepts, such as men's beliefs about how men should act, and they affect what they do, albeit always dependent on context. Thus the attribution of causation does not require essentialism or determinism.

Some anti-essentialists may see the attribution of causal powers and susceptibilities to particular *objects* as essentialist, and instead argue that what an object is and does always depends purely on the webs of relationships in which it is embedded. Of course, many phenomena (and all social phenomena) are strongly influenced by the relations into which they enter, but to deny that their behaviour has *nothing* to do with internal properties is an extraordinary dogma. For example, the nature of an industrial region has much to do with the relationships between people and technologies. But if we were to replace the people with lumps of rock, everything would be different. Causes of change can be both internal and external. Thus the capabilities of adults derive from a long process of socialization and evolution of internal and external powers. For this reason we need to keep an open mind when pursuing empirical research.

Although essentialism, properly understood, does not entail determinism, there are problems with it, especially in social science. One is that objects are represented as having not merely a kind of structural integrity, but as things which can never evolve into new, fundamentally different objects, with variation being limited to accidental properties. Thus, capitalism might be assumed to remain constant in its essential elements (capital–labour relations, capital accumulation, etc.) while undergoing change only in accidental or contingent matters. This seems unnecessarily restrictive, implying that the particular concrete forms capitalism takes are only superficial variations. There might indeed be variants of capitalism, such as British or French capitalism, which share the same basic structures, but some variants may involve more radical changes, for example, the development of worker-owned firms producing in competition with each other for markets (Sayer 1995).

Here it is important not to worry about the metaphysics of essences but to ask questions like: How deeply is capitalism changed by the replacement of individual capitalists by corporate forms of ownership? Does the alleged rise of the 'knowledge-based economy' change everything about capitalism or very little? Could capitalism survive a radical change in the gendering of its labour force? In developing answers to such questions we refine our understanding of the powers and susceptibilities of our objects of study, of what is necessary for them to be or do a particular thing, and what is not. And we get beyond the crude position of essentialism which obliges us to assume that there will always be a sharp difference between those circumstances which change an object into something completely different and those which change it superficially. For capitalism, I think it is better to be open to the possibility of gradual change, and in its fundamentals as well as its particular contingent forms. It is not a self-sufficient system and it always requires forms of non-capitalist organization, such as state regulation and households to function, and these vary considerably, thus helping to produce the significant variation that we see among capitalist societies.[1]

While we should reject both determinism and ideas of social phenomena having fixed, invariant essences, it remains the case that at any particular time, what an object (including any person or institution) can do is constrained and enabled by their make up and the relations in which it is embedded. This is simply the idea that we make our history but not in the circumstances, or with the materials, of our own choosing.

Gibson-Graham argue for what they term an anti-essentialist approach to capitalism (Gibson-Graham 1996). However, in their empirical work, as in their recent assessment of the Mondragon co-operatives in Spain (Gibson-Graham 2003), they try to identify constraints and affordances in any given situation, rightly recognizing that the reproduction or transformation of structures is influenced by ethical-political decisions about how to do things, and that action is constrained and enabled by existing structures (e.g., by growth of demand for the products the co-operatives sell, by links to other co-operatives and banks). Although they refer to anti-essentialist, post-structuralist theory at the start, they are really trying to identify the causal powers and susceptibilities (though they do not use this terminology) of the particular co-operatives in question.

So while the concept of essence seems well-suited to certain physical phenomena like minerals, and is not vulnerable to the accusation of determinism, it is less useful even for those social phenomena which are relatively durable and have structural integrity, for these are open to transformation in their fundamentals – not surprisingly, since they depend for their reproduction on skilled, reflexive agents. For these objects, we either have to adapt the concept of essence so it allows such variation in essential features, or dispense with it, and just talk about the particular properties (causal powers and susceptibilities) of particular cases. This enables us to discover what room there is for manoeuvre and hence avoid both political fatalism ('what is, is what must be') and political voluntarism ('anything and everything can be changed now regardless of its current nature'). This is fundamental to the relation between politics and method. If, on the other hand, we imagine either that there is no causation or that we can just act and make things happen differently regardless of the constraints and affordances of the situation, that is to invite failure in both explanation and politics.

Anti-essentialism is bolstered by a normative concern. It appears to liberate those whose oppression has been legitimized by being represented as naturally-grounded, for example in the case of racism by the kind of spurious naturalism that claims racial differences in ability. However, if we respond to such tendencies by denying that people have any particular natural properties as human beings, and assume these properties are no more than cultural or discursive constructions, we lose critical purchase on any oppressive exercise of power, particularly through torture, mutilation or abuse. This is disastrous for emancipatory movements. As Kate Soper (1995: 23) argues in relation to Foucauldian anti-essentialism:

> denying that there are any instincts, needs, pleasures or sensations which are not simply the effects of culture but impose their own conditions upon its 'constructions', then it is difficult to see what sense we can make of the notion of feminist reclamations of the body or selfhood from the distorting and repressive representations to which they have been culturally subjugated.

Certainly, there are dangers of identifying local and historically-specific characteristics as universal, and of failing to take seriously the remarkable variety of cultural forms that shape people deeply. In response to the treatment of local variants as universal or as the norm, and the common tendency to naturalize contingent historical forms of domination, it is tempting to reject any notion of human nature. Human beings are indeed extraordinarily diverse, but the question is what enables them to exhibit such variety? Humans can be variously culturally shaped, but not just anything can be culturally shaped. A lump of rock cannot take different cultural forms (it may be externally construed in different culturally mediated ways, and used in various ways, but granite doesn't change its nature when we think about it differently, any more than the earth changed shape when we decided it was round rather than flat.) For an entity to be shaped in a particular form (e.g., by culture) it must be the kind of thing which is susceptible to such shaping, that is, it must have (or have acquired) the affordances and resistances which allow such shaping.

Moreover, in line with Soper's point, we need to identify the capacities of humans for flourishing and suffering, enabling critiques of not just economic theories but economic *practices* affecting people's well-being (*Feminist Economics* 2003; Nussbaum 2000; Sen 1999). To be sure, there are many different forms of flourishing, and different cultures provide different conceptions of what constitutes flourishing, but not just anything can be passed off as it. If we were to insist that it was purely culturally relative, then we would have no warrant for using terms like 'oppression' (Nussbaum 1999). Thus anti-essentialism's understandable and admirable desire to do social science in a way which illuminates possibilities for emancipatory change could back-fire by making suffering and flourishing unintelligible (because they are merely products of discursive construction and nothing to do with our natures). I now want to discuss the nature of normative thought more broadly.

Normativity, politics and social science

When *Politics and Method* was published in 1985, the role of politics in geography and related social sciences was as a shared background concern. It informed researchers' relation to those they studied and the theories they chose, but they did not develop normative arguments about what was good or bad, progressive or regressive. Since then, if anything, the emancipatory hopes of critical social science have moderated, because of both the fear of imposing views on research subjects and increased suspicion of normative arguments. Although there is still some kind of radical geography community, the radicalism has become more muted. In radical economic geography a focus on how economic matters affect well-being and oppression tends to take second place to studying the latest capitalist developments or enthusiasms, such as the knowledge-based economy. Such studies may usefully expose certain misconceptions about their objects, but they do not necessarily address the general implications of contemporary economic organization for well-being. At best they leave the reader to assess these unassisted.

The connection between economy and well-being is not arbitrary. It is easy to forget that the point of economic activity is to enable people to live well, though under capitalism, this means–ends relationship is inverted so that 'the economy' (i.e. excluding the household economy) is something we are expected to serve, something which operates not to secure economic well-being but to make profit. Consequently, people may feel they live to work rather than vice versa. A purely 'engineering' focus on economic matters (Sen 1999), treating the contingent forms of economic relations as manifestations of universal deterministic laws, fails to challenge this extraordinary inversion which dominates modern life.

Like any social relations, economic relationships have a moral dimension in the sense that they presuppose understandings (sometimes contested) of rights and responsibilities and acceptable forms of behaviour towards others (Sayer 2000b). Thus, employment relations, exchange relations and credit relations could not exist without such norms. The welfare state formally defines certain economic responsibilities and entitlements and provides a corresponding bureaucratic system of taxation and benefits for linking them. Often, as in the case of property rights, norms may be imposed rather than freely and democratically arrived at, but they still have a normative aspect. Power, interests and habit are important too, but we couldn't function as social actors if we ignored these normative understandings. Sometimes economic power is exercised by taking advantage of workers' moral commitments, for example, their feeling of not wanting to let customers, workmates and families down. In addition, unintended consequences of economic actions also have ethical implications, ones which often play out in complex, tangled geographies.

These ethical dimensions and implications of economic activities *matter* to people, and not only in terms of income and wealth but in terms of the qualitative experience of the relationships themselves: whether they allow autonomy and responsibility, respect or lack of respect, tedium or fulfilment, exploitation, or the possibility of collective action. They affect mental and physical well-being. However, developing a critique of the subordination of life to the demands of capitalist economy is hindered by certain features of contemporary social science which inhibit both normative reasoning and recognition of the normative character of social life itself, as I'll now try to explain.

Up to the end of the eighteenth century normative and positive thought in the social sciences were fused, but since then they have been progressively separated, so that now, unless social scientists happen to study philosophy, they gain no training in normative thought, which also tends to make them dismissive of it. There has not only been an attempted expulsion of values from science, but a less-noticed expulsion of science or reason from values, so that the latter have widely come to be regarded as 'merely subjective' or beyond the scope of rational deliberation.

The divorce reflects changes in society, particularly the rise of liberal individualism and the related rise of markets (Poole 1991). One of the distinctive features of markets is that they do not generally require actors to justify their decisions: their money does the talking. Questions of the good are replaced by what will sell or can be afforded. Not surprisingly, during the same period, the language of political economy became less moralistic, so that instead of the rich moral economic vocabulary of self-interest, greed, envy, vanity, benevolence, compassion, profligacy, prudence and virtue, we have bland 'interest' and 'utility' (O'Neill 2004). Instead of a study of how economic relations fitted into the wider order of society and of the ethical implications of such relations, political economy developed a narrow engineering focus which unreflexively reproduced the capitalist 'de-valuation' of economic behaviour. Hence, also the rise of 'subjective' theories of value, interpreting valuations of and commitments to particular ways of life as mere individual preferences, which is exactly how they appear in markets.

This expulsion of values from science and reason from values has produced not only generations of social scientists who are both ill-practised in normative reasoning and dismissive of the very idea, but who tend unknowingly to project their 'de-normativized' orientation on to the world, thus producing an alienated and alienating social science which struggles to relate to everyday experience and why anything matters to people (Manent 1998; Sayer 2004).

These tendencies even apply to radical or critical social science. In these one can get away with using at least some evaluative terms like 'oppression', 'exploitation' and 'racist', provided one uses them in conventional ways acceptable to the readership. However, if one goes further and describes institutions or practices as 'unethical' and provides an argument about why certain things are bad, in a way which indicates the standpoint or conception of the good from which the critique is being made, one is liable to be challenged, and not only by conservatives but increasingly by radicals themselves. This is because it has become common under the influence of post-structuralism to regard normative argument as authoritarian,[2] rather than as merely an invitation to join in the discussion. Criticism is increasingly giving way to a universal scepticism which undermines criticism by refusing any possible standpoints from which judgements can be made.

Ironically, while many radicals claim to refuse the fact–value dualism, they also often regard normative argument as 'merely subjective' and as having nothing to do with the properties of states of the world. In so doing they reassert the distinction they claim to reject. However, the aspects of life which are most important to radical politics – whether people are flourishing or suffering, whether they are able to meet their needs – are precisely the ones that elude classification either as purely positive or purely normative. Addressing such issues involves thought which is simultaneously positive (world-guided) and normative (world-guiding). When we say 'unemployment tends to cause suffering', we are not merely 'emoting' or expressing 'subjective views', but saying something about the world. Nor are we simply providing a purely positive description because it implies the situation is bad. We may try to find more neutral terms, but if we do, one of two things will happen: either we will mislead by euphemizing, allowing others to fail to understand that suffering is involved (in which case the description will be positively as well as normatively deficient), or the new words will come to take on a similar evaluative load to the old ones. Similarly, when we refer to a need, such as the need of the young and the sick for care, our ideas are simultaneously positive (in recognizing the need) and normative in suggesting that the world should be changed to satisfy that need. Of course any such claims will be fallible, but that applies to all knowledge, and hence 'cancels all the way through'. Moreover, fallibility presupposes that knowledge claims are about something independent and hence about which they may be wrong. Interestingly, these matters which cannot be categorized as either positive or normative are not trivial curiosities but the things which are most important to our lives.[3]

Having strong normative views about the things we study does not necessarily lead us to misrepresent them, for it is possible to acknowledge unpalatable facts. Such values may help one notice things others miss. For example, feminist values help feminists to identify aspects of social reality previously overlooked. Values should therefore not be seen as a form of contamination threatening the development of scientific or adequate knowledge, but as deriving from claims about flourishing and suffering (Anderson 2004).

Radical geography's original largely Marxist inspiration meant that it inherited Marxism's uncomfortable relation to normativity and ethics. Its combination of unacknowledged ethical critique and a tendency to see science as counterposed to ethics affirmed rather than challenged the fact–value dualism. As in modern social science more generally, its scepticism about the status of normative ethical argument was complemented by a tendency to disregard the subjective normative experience of actors, as evaluative beings, continually assessing how they and others are being treated. Politics was apparently purely a matter of power and interests, things which supposedly could be studied scientifically, in contrast to the soft-headed pieties of ethics. The trouble with this is that politics without ethics is dangerously directionless, as open to fascism as to anything progressive, because it cannot offer any justification for its programmes or understand the ethical quality of the everyday life which it seeks to change. Without a conception of the good or well-being we cannot say why any particular form of power is progressive or reactionary. Radicalism without ethics is a contradiction in terms.

Feminist theory has developed a much richer normative critique of gendered economic orders. One possible reason is that the interpersonal and uncommodified nature of an important subset of gendered economic relationships makes their ethical implications much clearer. In discussing issues like care, dependence and autonomy, and how these affect well-being, they contribute to a general conception of the good (e.g., Sevenhuijsen 1998; Tronto 1994). In other words, feminism takes questions of what forms of everyday life are good (conducive to flourishing) or bad (conducive to suffering) much more seriously than Marxism, and arguably this enables it to contribute more directly and constructively to theoretical and political thinking about normative issues.[4]

Flourishing and suffering are not merely subjective matters in the sense of being no more than products of our imaginations, but objective[5] states which we attempt (fallibly/'subjectively') to identify and comprehend. The oppressed and despised cannot just turn their suffering into flourishing by an act of will. If flourishing and suffering are no more than arbitrary attributions of our imaginations, then any kind of radicalism is dead in the water, for any problem could be remedied simply thinking about it in a different way (Nussbaum 2000; Sayer 2004).

Conclusion

Despite the extent of the debate, 'essentialism' is not a well-understood concept, but then nor is it particularly appropriate for social science. Neither essentialism nor anti-essentialism provides a useful entry point into matters of causation, and difference or variation. In the limited space available I have tried to provide a few indications of how they need deconstructing. We need an approach to economic geography and related subjects which allows us to analyze social phenomena ranging from the durable and structural through to the ephemeral in a way which takes into account their diverse causal powers and susceptibilities (what they can and cannot do), what is necessary and what is contingent in relation to any object (what they must have or merely can have), bearing in mind: (1) that many objects exhibit considerable variety and hence have varied properties; and (2) people have the power to be reflexive and hence to change both their circumstances and themselves, albeit, at any given time, in accordance with the powers and susceptibilities that they, and other objects, currently possess. Many of the objects that interest us may be networks of relations, but these too have particular properties, whether active or latent, which must be explained, so that we can establish what it is about them that enables them to do certain things and not others. Social phenomena are also influenced by discourses and actors' concepts and these need to be interpreted too and examined for their causal impact.

While these are standard critical-realist ideas I have tried to go further and discuss the normative, political dimension of social thought, which radical geography, like most critical social science, tends to be wary of elucidating, probably because normative thought is cast as the opposite of science, as a matter of individual subjective opinion. Ironically, the things we care about most – what is good, what is bad – are treated as beyond the scope of reason, thus undercutting any grounds for radicalism or critique. While normative arguments are very difficult to defend, I have argued that we will misunderstand them if we imagine them to be merely subjective, or mere discursive constructions. Rather, I have argued that they concern our nature as beings capable of flourishing and suffering, and having a range of needs. A politics which ignores this and is purely about power could be utterly oppressive. If we reject such claims about needs and flourishing on the grounds that they appear 'essentialist', we will render normative thought, including radicalism, indefensible. Rather, we should seek to identify what forms of flourishing and suffering are present in the situations we study, and what sustains them and what changes them.

Acknowledgement

This chapter was written with the support of an ESRC fellowship.

NOTES

1 There is also a practical reason for preferring this strategy – the lack of a clear successor system.

2 I have found that if I go beyond the usual degree of normativity people challenge me to say where my critique 'hails from'. It's a perfectly reasonable question, though I don't see why it shouldn't be asked of those who use more conventionally limited evaluative descriptions. My answer would be that it hails from a qualified ethical naturalism, similar to that defended by Nussbaum (1999, 2004).

3 Nor can they be dismissed by reference to the alleged 'naturalistic fallacy' of claiming that ought follows logically from is, for logic concerns the relationship between statements, whereas the relationship between needs and their satisfaction, or suffering and its relief, are relations of states or processes, not logic. A starving person doesn't need a logical warrant for getting food.

4 However, as Sevenhuijsen notes, it is often wary of the term 'moral', because of the tendency of oppression to be masked as moral. Some may prefer 'justice' to morality, but that tends to be associated with a narrow conception of the good which fits better with liberalism and its disregard of relations of care, than feminism.

5 It's important here to avoid a common confusion arising from slides between three different, logically independent, meanings of 'objective': (1) 'value free'; (2) 'statements believed to be true'; (3) 'pertaining to objects themselves, regardless of how they are understood'. It is only (3) that I mean here (Sayer 2000b, see also Collier 2003).

8 CULTIVATING SUBJECTS FOR A COMMUNITY ECONOMY

J. K. Gibson-Graham

In the local action research projects in which we are awakening and implementing different economic possibilities, we are engaging in 'a politics of the subject' (Gibson–Graham 2006). Minimally this means producing something beyond discursively enabled shifts in identity, taking into account the sensational and gravitational experience of embodiment. If to change ourselves is to change our worlds, and if that relationship is reciprocal, then the project of history-making is never distant but right here, on the borders of our sensing, thinking, feeling, moving bodies.

Wary of producing a private language in which to conceive diverse economies, we wanted to converse with people willing to entertain the idea of a different economy. In our research interactions in Australia, the USA and the Philippines we hoped to *speak* and *hear* richer, more vibrant economic dialects, to explore and develop our rudimentary language of economic difference, to construct alternative economic representations, to cultivate subjects for a community economy,[1] and ultimately to build a linguistic and practical community around new economic projects and possibilities.

This chapter is about projects undertaken in the Latrobe Valley east of Melbourne in Australia and the Pioneer Valley of Western Massachusetts in the USA. Much of that work was performed by paid 'community researchers' with little prior experience of research. In the Latrobe Valley Community Partnering Project three academic researchers recruited and trained five community researchers (CRs) drawn from groups marginal to the mainstream vision of an economically and socially 'developed' region – retrenched electricity workers, unemployed young people, and single parents. In the Pioneer Valley Rethinking Economy Project twelve academic researchers (both volunteer and paid) trained seventeen CRs of different ages, colours, genders and ethnicities to conduct interviews with friends and acquaintances who participate in non–capitalist sectors.

Participatory action research in its traditional guise attempts to break down the power differential between the researched and the socially powerful by enabling the oppressed to become researchers of their own

TABLE 8.1   **Methodological steps of the action research projects**

Step 1. Tracking practical economic knowledge in each region by convening focus groups with key participants in the local economic development conversation – planners, governmental officials, labour and business leaders – and key participants in local community development – NGO workers, social workers, service providers, religious and community leaders.

Step 2. Acknowledging the positives and negatives of existing representations by documenting the stereotypes of the region and its people within the local economy. These findings are presented in visual form to highlight two different paths for mobilization – the 'half empty' or 'half full' approaches.

Step 3. Generating new representations via a 'community economic audit' of non-mainstream or undervalued economic activities. The audit is conducted in collaboration with community researchers drawn from the alternative economic sector or from economically marginal or undervalued populations (the unemployed, housewives/husbands, recent immigrants, welfare recipients, retirees, youth, etc.). It delineates the range of economic activities in the region and provides a setting in which to reframe individual economic identities.

Step 4. Documenting inspiring examples of community economic enterprises or activities. The purpose of these short case studies is to illustrate the variety of ways of populating the community economic landscape and provoke the desire to be a part of it.

Step 5. Creating spaces of identification in which community and academic researchers, community members, and local institutions begin to imagine becoming actors in the community economy, identifying with various economic subject positions. These spaces include field trips, social events, displays, community brainstorming workshops and specialized 'how to' workshops.

Step 6. Acting on identifications within the community economy. In this step heterogeneous groups begin to build enterprises, organizations and practices that construct a community economy.

circumstances (Freire 1972). From our post-structuralist position, we don't accept that power can be banished. There are always *multiple* power differentials at play. Our projects aimed at mobilizing desire for non-capitalist becomings. That desire stirs and is activated in embodied interactions in which power circulates unevenly yet productively across different registers.

In what follows we draw upon both projects to reflect on the complex and challenging process of cultivating novel economic subjects. Our reflections are organized under five headings that loosely track the research steps in Table 8.1:

- discursive destabilization;
- resistance and non-recognition;
- reframing and re-narrativizing;
- producing positive affect;
- spaces of identification and ethical openings to communality.

In each section we highlight moments in which the embodied and the discursive interact to create new ethical possibilities, and that signal to us the emergence of subjects of the community economy.

Discursive destabilization and its unexpected effects

Initial focus groups attempted to access and dislocate the local countenance of Economy. In the Pioneer Valley, planners and business and community leaders were asked to speak about the strengths and weaknesses of the regional economy. Familiar stories emerged, couched within the anxiety-ridden discourse of regional development that justified intervention. The prescription: attract 'good' jobs to the region by recruiting major capitalist employers via subsidies and other inducements.

But the conversation took an unsettling turn. Some participants noted that economic development requires a suitably educated and acculturated labour force, while others said the two-earner family, whether wealthy or impoverished, leaves no one at home to raise the children. Discussion circled around the unspoken fear that successful economic development might bring social failure. Participants harboured a concern that a capitalist economy (no matter how developed) is insufficient to the task of sustaining a community – raising its children, reproducing its sociality. This concern surfaced at several moments, signalling a discursive non-closure, an unsettling of capitalism's subjects. Over the course of the project we would come to see these moments as emotional openings for a counter-discourse of 'the household-based neighbourhood economy' (Community Economies Collective 2001).

The focus groups held in the Latrobe Valley were also with business and community leaders. When asked to talk about the social and economic changes transforming the Valley over the last decade, they produced a well-rehearsed story centred upon dynamics in the formal economy – privatization of the State Electricity Commission (SEC) and the pressures of globalization. Words such as 'victimization', 'disappointment', 'pawns', and 'powerless' anchored this narrative of negativity, and the moods of the speakers ranged from energetic anger to depressed resignation.

But when they were later asked to consider the strengths of the region and its capacities for change, an unmatched set of stories emerged, conveyed in that halting manner of speech in which new thoughts are expressed. Participants spoke of artistic ingenuity and enterprise, of contributions made by migrants and the intellectually challenged, of the potential of the unemployed. Moods began to lighten, and expressions of surprise and curiosity displaced dour agreement.

To our surprise and pleasure, the stories began to map the contours of a relatively invisible diverse economy. Not simply abandoned by capital, the region was populated by community-based economic alternatives. At the end of the session, one participant noted the shift that had occurred in his own understanding:

The information that I've gained just from hearing everybody talk ... it's been absolutely precious. And it hasn't come about as a consequence of some bureaucracy wanting to solve problems but rather as we are pawns in another exercise [i.e., our research project]. I'm actually going away from here with more than I came with. (Local government official)

Here were economic development practitioners, business people, union officials and local government functionaries dropping their old habits, speculating about alternatives. We felt heartened but uncertain about the next phase. Could the momentary relinquishing of established identities be prolonged so that new subjects might emerge? Would we move with them towards a new understanding and practice of economy?

Encountering resistance and non-recognition

As part of a qualitative assessment of life in the Latrobe Valley, we asked each community researcher to create a photo essay that told 'their story' alongside that of their friends and acquaintances. We were both surprised and confronted by the result. These photo essays were used at various points by the community researchers in their conversations, and within that forum the psychic difficulties of relinquishing established economic identities came to the fore.

The essays provide insights into the very different life worlds of the retrenched worker, the young unemployed, and the single parent. In 'Jock's Story', the ghost footprints of an industrial worker trudge through scenes of abandonment – Jock stands waiting for his mates at the vandalized bus stop where electricity workers once boarded the SEC work bus (see Figure 8.1). There are images of derelict industrial buildings, abandoned mining equipment, a vacated workers' parking lot at the power station, pawn shops and boarded-up premises in the main street of Morwell. The retrenched workers convey their anger at both the empty promises of growth with privatization, and an electricity industry still producing coal and power while living breathing labour is nowhere to be seen.

'The Young Latrobe Valley' photo essay evinces a similar abandonment among the Valley's youth who have nothing to do except to vandalize buildings, drink, smoke, gamble, eat at McDonalds, visit Centrelink and the courts. There is an overwhelming feeling of frustration, of having been betrayed. Included in the poster are examples of the predominant media depictions of the Latrobe Valley as blighted and desperate: 'The Valley of the Dole' (Tippett 1997), the 'Valley of Despair' (Shaw and Munro 2001). Above all there is the insistent anxious questioning about this generation's future and an implicit mourning for its lost former identity.

In both valleys, we found that unlike unemployed middle-aged males and youth, low-income, female single parents refused the negative

FIGURE 8.1 Jock's story

identities offered by the dominant discourse of economy. AnnaMarie, a feminist planner and academic researcher, was all too aware of the mainstream representations of single-parent households in the inner-city neighbourhoods of South Holyoke in the Pioneer Valley:

> Planners portray them as economically deficient, or even as economically depleting if the women are receiving government assistance. ... The underlying mainstream assumption about low-income communities is that they are somehow outside 'the economy', or are lacking an economy, and so an economy needs to be developed there. In our Rethinking Economy project, we assume that there is already an economy in low-income communities. In fact, we assume multiple and diverse forms of economic activity taking place – both legal and illegal, for pay and not for pay, capitalist and non-capitalist, involving barter, gift, theft and market transactions. Furthermore, we assume that the households in these communities constitute important sites of (largely unpaid) economic activity. Where the mainstream sees absence or emptiness, we see presence and fullness.
>
> In the focus group I conducted with neighbourhood women, I did not encounter empty, deficient or depleted subjects. The women were intellectually lively and full of playfulness.

While the men and youth were disoriented, lacking identities in the mainstream economy, the women of both valleys were not captured by dominant forms of economic subjection. In their partially disinterpellated and disidentified state, they were visible as potential spaces of cultivation.

Despite the resistance and negativity, we were encouraged to pursue the project of destabilizing existing identities, prompting new identifications and cultivating different desires and capacities. We began by reframing the economy and economic roles and re-narrativizing economic experience.

Reframing and re-narrativizing

One of the goals of the action research projects was to flesh out through a community inventory a diverse economy in which capitalist enterprises, formal wage labour and market transactions occupy only the visible tip of the economic iceberg. By emphasizing activities that are often ignored, we hoped to refigure the identity and capacities of the regional economy. Further, by recognizing the particularity of people's economic involvements, including their multiple economic roles (capitalist and non-capitalist), we reframed an individual's identities and capacities. We undertook slightly different reframing exercises at each site but central to each was the involvement of community researchers.

Over a period of weeks in the Latrobe Valley the academic research team facilitated an intensive training of the CRs. Initially, we discussed different ways of thinking about economy, including the representation of a diverse economy (see Gibson-Graham 2006: Chapter 4). For illustration and inspiration we visited community-based enterprises and craft networks in the Valley. None of the CRs knew about these 'alternative' economic organizations. But after this exposure their perceptions of the Valley and its potentialities changed.

As a result, the CRs brought a new fullness into the representation of identity. No longer were subjects defined as lacking or victimized; instead they were invited to see themselves as skilful and competent. Key to this process of self-reinvention was the Asset-Based Community Development (ABCD) method. Pioneered by John Kretzmann and John McKnight (1993) in their work with community activists and organizers from inner-city neighbourhoods of large North American cities, ABCD is a simple yet powerful tool for reframing community identities and re-narrativizing community development trajectories. The method takes the familiar image of a half-empty glass, likening it to those communities that self-define by what they lack and therefore 'need'. Such needs-based portrayal invites solutions from the outside – grants and assistance from government and non-government agencies. It perpetuates dependence, consigning community members to write grant proposals rather than acting as self-managers. Kretzmann and McKnight suggest another path: a half-full vision emphasizing assets and capacities of residents, associations, and local institutions. They argue that communities are built on their assets, not on their needs.

Generating new representations of the Latrobe Valley involved marrying the ABCD method with the language of the diverse economy. The community researchers began with group interviews to identify people's assets and capacities for what we termed a Portrait of Gifts, that is, the qualitative information 'outside the dominant story' critical for creating new representations of the Latrobe Valley. The aim was not to produce a complete inventory of skills but to shift people's self-perceptions. Most important was the process of working together to complete a joint re-portrayal. People invariably surprised themselves with the extent of their capacities, learned new things about each other, and found common areas of interest. In place of depictions of the Valley as distressed and dysfunctional, the Portrait of Gifts offered a shared representation of a 'caring community', and a glimpse of a new kind of economic subject concerned for and connected to others. With this alternative story of the Valley, the community researchers could pursue the possibilities of residents playing new roles, rescripting themselves as active makers of a community economy.

A similar process of reframing and re-narrativizing took place in the Rethinking Economy Project in the Pioneer Valley. Initially, the primary economic identification of the CRs was with capitalism. They were actual or potential workers, entrepreneurs, consumers, investors, and their politics was structured by antagonism or admiration for capitalism. Capitalism was the master signifier, organizing economic space via identification and desire. The challenge was to dislocate this formation and to create the emergence of non-capitalist forms of economic subjectivity. During the training, a drawing of an iceberg was used to portray the economy, with the capitalist economy above the water line, and a diverse, huge, largely non-capitalist economy below it. We invited the CRs to collaborate with us to know and speak about the economy under the water line.

Over the next few weeks, each CR conducted six two-hour interviews with residents about their daily activities. At the weekend debriefing retreat we asked them to situate activities from their interviews on a four-cell matrix taped to the wall. Predictably, the examples were scattered among the for-market/non-market, paid/unpaid boxes. People were impressed by the diversity, but what also emerged unexpectedly was the extent of an economy of generosity, overflowing with gifts of goods, money and labour. At the end of the day we asked the researchers to affix red stickers to any activity they considered capitalist, and to the amazement of everyone only four of the more than 50 entries received stickers.

Coming to a new language and vision of economy was an affirmation not only of difference but also of economic capacity. People in our research encountered themselves differently – not as waiting for capitalism to give them their places in the economy but as actively constructing their economic lives, on a daily basis, in a range of non-capitalist practices and institutions. Development possibilities were transformed by giving them different starting places in an already viable diverse economy.

Producing and sustaining positive affect

Becoming a different economic subject is not an easy or sudden process. It is not so much about seeing and knowing as it is about feeling and doing. When we began working in the Latrobe Valley with those who were marginalized by economic restructuring we often encountered hostility and anger, anchored in a deep sense of powerlessness. Introducing our project of economic resignification reactivated the trauma of retrenchment (especially for men in their 40s and 50s who had found it impossible to secure alternative employment) and reinforced the bleak future envisioned by young unemployed people. Rather than counting on language and discourse to release and redirect affect, we found we needed to directly address embodied and emotional practices of being; that is, to work with the emotions that blocked connection and receptivity to change.

Much of the animosity was directed to the CRs. The project was an initiative of Monash University (which has a regional campus in the Valley) and the Latrobe Shire Council, so the CRs were initially identified with these institutions. Many in the Valley saw these powerful regional actors as stand-ins for the SEC, directing their anger towards their representatives. Yvonne, one of the community researchers, became quite skilled at dealing with negativity. Reporting on her somewhat difficult conversation with Wayne, for example, she says:

> [I found out that] he is very good with his hands and knows a bit about cars. I asked, hypothetically, if there were a group of single parents interested in learning about car maintenance, and if I could arrange a venue and possible tools, would he be interested in sharing his skills and knowledge? 'Yeah, I'd do that, no worries,' he said. I asked him would he expect to be paid for his time. 'No, I wouldn't do it for money,' he replied. I asked, 'So do you think you'd get anything out of it yourself?' 'Yeah, I suppose I'd get some satisfaction out of it 'cause I like to help people like yourself.' So I really tried to turn it around and have him answer or resolve his own questions and issues.

Yvonne's strategy involves moving away from narratives that trigger anger and towards the pleasures associated with a different way of being. Through her patient questioning, Wayne's attention is shifted away from a narrative of impotence and victimization and towards a scenario that positions him as skilful and giving, endowed with an identity in a community economy.

Varela writes that 'We are already other-directed even at our most negative, and we already feel warmth toward some people such as family and friends ...' (1992: 66–7). This 'being-in-common' (Nancy 1991) can be extended to create new practical knowledges for a community economy (Varela 1992). Fleeting ethical moments such as that related above might appear to represent very minor shifts in the macro-political scheme of things, but they are requisite for the political process of enacting a different regional economy, where individuals from different backgrounds move beyond fear, anger and hatred to interact in inventive and productive ways.

Kara, one of the Pioneer Valley CRs, exemplifies both Varela's notions of extending familiar feelings to unfamiliar situations and self-cultivation in fostering and stabilizing positive emotions. Kara was initially highly resistant to the project's goal of bringing mainstream and marginalized economic actors into conversation and collaboration. She thought the marginalized would be subsumed to the agenda of the powerful. Mainstream people, she believed, were emissaries of the State, Capitalism, and Power, the unholy trinity, from which she was hoping and indeed planning to escape. Throughout the weekend training for community researchers she vociferously communicated her antipathy. But at the last moment, during the evaluation of the training, she suddenly saw the possibility of productive engagement with those (threateningly) different from her: 'I don't want to be so us–them,' she said, 'or to live in a world that is set up that way, emotionally and politically'. Abruptly the mainstream types were rehumanized, and the possibility of co-operating became a political opportunity. It was as though in that very moment Kara was working on herself 'to attenuate the amygdalic panic that often arises when you encounter ... identities' that call the naturalness or sufficiency of one's own identity into question (Connolly 1999: 36). By engaging in a 'selective desanctification of elements in [her] identity' (146) – in this case, a highly charged oppositional stance with respect to power understood as domination – Kara opened herself generously: to the humanity of others, to the possibility of being other than she was, to participating with those most different from her. By recalling her principles and returning to her store of skills and capacities, she called up feelings of respect and connection, extending them to strangers and committing publicly to a positive emotional stance.

Creating spaces for new identifications and ethical openings to communality

As a way of building upon the dreams that emerged from initial conversations in the Latrobe Valley, we organized a range of events that drew people together in situations where crazy ideas and schemes could be freely thrown around without the pressure of a formal meeting. Pizza-making and scone-baking events were particularly successful in getting people to meet, overcome the stultifications of shyness, begin to listen to one another, and build and transmit excitement. Leanne described a moment of understanding prompted by these gatherings:

> I guess the crunch came when Yvonne was doing her food events and things like that. And just the mingling and talking to people, to me that was like the breakthrough of … this is what it's about, it's working with other people and listening to other people and getting that opportunity to listen to their dreams and things like that.

She went on to consider how she herself had changed over the course of the project:

> If I give myself the time, I can listen to anyone. I had only ever dealt with people over the counter [before involvement in the project] – with commercial transactions. I'm not as critical as I was. Working with people from various backgrounds and abilities – I'm more tolerant. I've learnt to see the good in people. I had always been taught to be cautious and careful of people. My dad always used to say, 'the only friend you've got is yourself'. But the project is a place where you can relax and take people as they come. It offers the security to trust people.

Linda Singer suggests that we understand community 'as the call of something other than presence' (1991: 125), the call to becoming. And the capacity for becoming is the talent we were perhaps most actively fostering – through individuals opening to one another and to the inescapable fact of their 'own existence as possibility or potentiality' (Agamben 1993: 43). Almost every meeting and event associated with the projects stimulated desires for alternative ways to be, and each of these desires operated as a contagion. In the face of this enthusiasm, however, there was still uncertainty about what a community enterprise actually was and how to build one. Many people had never been involved in community organizations or other types of non-mainstream work.

In the Latrobe Valley we organized 'how to' workshops in response to ideas that arose during earlier brainstorming. At one of these sessions, Gil Freeman presented the story of CERES, a very successful community garden in inner-city Melbourne, and answered questions about its operations. As a result, two interested groups from the Valley travelled together on the bus to Melbourne. While CERES itself made a strong impression, the experience of the bus trips – of being cooped up together in a moving steel canister for hours at a time – produced an atmosphere of enchantment that became the life force of a nascent garden project. Very soon a 'mixed group' of retrenched workers, retirees, housewives, people of non-English-speaking backgrounds and unemployed youth obtained access to an abandoned caravan park in Morwell and began the Latrobe Valley Community and Environmental Garden (LTVCEG). Among LTVCEG participants there was a general feeling of hopefulness and astonishment that they had initiated an enterprise. A space opened up for relations between people who thought they had nothing in common and a community economy began to develop.

William Connolly identifies an 'ethos of engagement' as an aspect of a politics of becoming, where subjects are made anew through engaging with others (1999: 36). In the context of our projects this transformative process involved cultivating generosity in the place of hostility and suspicion and awakening a communal subjectivity. Partly as a result of our efforts, we evoked a faint but discernible yearning for a communal (non-capitalist) economy.

In the Pioneer Valley, our efforts were similarly inclined towards bodily practices of engagement as a supplement to the delights of words. We tried to make our conversations and gatherings entirely pleasurable (food was one of their main ingredients) and loose and light – not goal-oriented or tied to definitions and prescriptions of what an 'alternative' should be. Over the course of the project, without prompting, the CRs and their interviewees began to express practical curiosity (as opposed to moral certainty) about alternatives to capitalism. We decided to travel together to Cape Breton to attend the Festival of Community Economics, a conference on worker cooperatives. Ten of us piled into a UMass Geosciences Department van for the seven-day outing, including members of our university-based group, community researchers, activists and local NGO workers.

Talk at the conference focused almost entirely on co-operatives. By contrast, the van was a conversational plurispace, with Jim, a community researcher, as conservator of spaciousness. Jim offered a passionate tentativeness as his approach to life and to economic possibility. He became particularly worried that growing a community economy would become a monocultural practice. 'Are co-ops the only answer?' he asked, disturbed by the lack of alternatives (to cooperatives) discussed at the conference.

One afternoon on an outing, we found ourselves at a dead end in a bog that must be one of the wonders of Cape Breton. Jim became our guide and naturalist, introducing us to the miniature laurels, roses and irises that flowered there. Crowns low against the wind, roots seeking scarce soil for sustenance, these tiny ancient plants were fragile yet incredibly hardy. In a land both familiar and foreign, we stumbled as lost travellers upon a thriving yet barely visible ecosystem. What was this bog if not a ground portrait of a diverse community economy? We discovered what we were seeking in Cape Breton without knowing where to look for it.

Throwing ourselves together in the van, embarking on a trip without a clear or single vision of what the goal might be, we found our desires dispersed and differentiated yet activated by a quest for an alternative economy. The van became a travelling space of conversation and connection among different visions, projects, hopes, languages, experiences, and levels of involvement or commitment. By the end of the trip, we had produced several fantasies of cooperative enterprises and the social life they might enable, as a way of performing and acknowledging our temporary, satisfying collectivity. The affective stance of the van travellers was striking. There was no militant advocacy, no talk of struggle against a despised capitalism; co-ops were not the be-all and end-all, just an everyday (presumably problematic) place to nurture a community. The passionate tentativeness we first saw in Jim became the spirit of the van community. The trip itself became an acknowledgment of the ways that we are all already in a space of communality.

A postscript and invitation

The process of 'history-making' or 'disclosing new worlds' (Spinosa et al. 1997) entails new discourses, new identifications, exploratory acts of self-transformation, and the fixing of new 'dispositional patterns of desire' (Connolly 1995: 57). At any point in the history-making process individuals are caught between the dissatisfactions and disappointments of what they know and habitually desire and the satisfactions and surprises of what is new but hard to fully recognize and want. They require nourishment and encouragement from without to sustain acts of self-cultivation, to see changing selves as contributing to changing worlds. When our projects came to an end, one of the sources of that nourishment and encouragement was locally dissipated. The 'constitutive outside' that supported and connected these experiments in self-transformation was itself transformed. But it did not disappear entirely – its locus shifted from these specific action research projects in place to the overarching and ongoing project of the Community Economies Collective (www.communityeconomies.org). The goal of this collective is to create the discursive and institutional conditions for the emergence of 'community economies', transforming ourselves and our situations in the process. We invite others to join us in this potentially world-disclosing project.

Acknowledgments

Without the work of our frequent co-authors and constant collaborators, Jenny Cameron and Stephen Healy, none of the activities described in this chapter would have taken place. Without the 'editorial' work of Trevor Barnes, the chapter itself would not exist. We are forever indebted and very grateful to Jenny, Stephen, Trevor and to the many others who worked with us on the action research projects. Research in the Latrobe Valley was supported by grants from the Australian Housing and Urban Research Institute and the Australian Research Council (Grant Nos. A79703183 and C79927030) with further financial and in-kind support provided by the Latrobe City Council. Research in the Pioneer Valley was supported by a grant from the National Science Foundation (Grant No. BCS-9819138).

NOTE

1 'Community economy' is not a geographic or social commonality but an ethical space in which community is performed.

9 A PUBLIC LANGUAGE FOR ANALYZING THE CORPORATION

Phillip O'Neill

... the way in which people comprehend and make sense of the social world
has consequences for the direction and character of their action and inaction.
(Purvis and Hunt 1993: 474)

Economic knowledge is saturated with language. This chapter is about the
task of creating effective language that engages public audiences with
strong, well composed ideas. It builds on two thoughts, both about language
and its domains.

One thought is Virginia Woolf's: '... the art of writing has for backbone
some fierce attachment to an idea ... something believed in with conviction
or seen with precision and thus compelling words to its shape ...' (Virginia
Woolf 1938: 280–1). These words come from Woolf's deliberation on 'The
Modern Essay'. They compel us, as writers, to know, probe and capitalize on
the relationship between language and thinking, to be impatient with the
scope and detail of word selection and arrangement, language's shape.
Simultaneously, and through this interrogation of language, we must pursue
and construct an idea, a worthwhile idea, and elaborate on it with fierce
attachment. Woolf's words are a pronouncement, a call to arms, rather than
an argument. Yet eight decades on, they speak to an unresolved tension
about the nature of economic knowledge arising from its disturbance by
post-structuralism. This chapter also addresses this disturbance.

The second thought comes from George Orwell. In 'Why I Write'
(2004), Orwell angrily noted that the times in which he lived drove him to
write for political purposes even though he would have preferred to
explore the simple joys of writing. Edward Said similarly elevated a political duty on behalf of Palestinians and others in his writings. The intellectual's role, he wrote, '... is to uncover and elucidate the contest, to challenge
and defeat both an imposed silence and the normalized quiet of unseen
power, wherever and whenever possible.' (2001: 31)

Language, then, has substantial political possibilities. As academics,
though, we have a reserve about public engagement, perhaps arising from
the academy's insistence on high standards of evidence, robust configurations of argument and a desire for durable explanation. Orwell and Said

would not want us to reject these but include them into public languages of engagement as scholar activists. Simply, language is the device we use to engage with our audience. Yet there is an uncanny public silence among us, given the import of what we say to each other in workplaces and at conferences. My suspicion is that we write largely for ourselves, with little expectation of encountering a substantial readership. Not surprisingly, there is a high level of disconnection between the academic author and the public audience. This gap needs bridging for two reasons: first, economic geography needs a public constituency for we have something worth saying; and, second, engaging this constituency broadens and sharpens our thinking.

So this chapter aims to build consciousness of the ways to construct formidable political-economy arguments. It uses examples from my own public engagements where I am variously a researcher, a 'public expert' and a regular newspaper columnist. The reader should note that these have involved commentaries on relatively minor events to regional audiences – so the stakes haven't been high on a national or global scale. Yet, perhaps, this gives me experiences and reflections easily shared. First, though, the disturbance of post-structuralism needs comment.

Towards a contemporary theory of knowledge

In *Politics and Method*, Massey and Meegan (1985b) opened a debate in economic geography about the relationship between research methods and the nature and purpose of knowledge. The debate was of its time, obviously, centred on the extent to which economic change in different places could be seen to be caused by common, underlying relations and processes; and the extent to which these could be revealed by both extensive and intensive research methodologies. Of most interest was the development of close study techniques to reunite theory, methodology and policy in research, seen to have been separated by positivist empiricism, especially involving quantitative techniques.

Two decades on, we have a different conception of knowledge and, therefore, of the research process and of political arguments and interventions. At the heart of this difference is a new understanding of the role played by language. Broadly, the shift in our understanding of the nature of knowledge is captured by the term 'post-structuralism'. Three processes drive the commitment in social theory to post-structuralism. The first is the evolution of post-structuralism from being a specific movement within linguistics concerned with the way text constructs existence, to an interdisciplinary coalition which places the idea of existence somewhere on a continuum between language-enabled and language-enacted life. The second is a crisis of faith in modernism as capable of delivering a higher-order social existence through objective knowledge-creation processes. The third is a refinement of post-structuralist insights and methods across the social sciences and an acceptance of the inscribed nature of human subjectivities, of reflexiveness in the creation and meaning of texts and of the role of language in the assembly and use of power. Together, these have transformed human geography. There remains, though, the important task of building a post-structuralist politics. Here I suggest three propositions about language and politics: that language is intrinsic to all processes of socio-institutional change; that language construction is inevitably a reflexive event; but that the political power of language comes from it having purpose.

Language is a structured device for socio-institutional change

Post-structuralism tells us that the composition of economic theory rests heavily on language as a sorting, explaining and suggestive instrument. A systematic treatment of the power of language in constructing knowledge is developed in Fairclough (1992, 1995, 2003), who argues for the inseparability of text, discursive practice and social practice. Now accepting that social practice is infused with language requires something beyond deconstruction: Fairclough stresses that language analysis is not just useful in social analysis but that social analysis is impossible without it: 'Discourse is a practice not just of representing the world, but of signifying the world, constituting and constructing the world in meaning' (Fairclough 1992: 64). As such, political and economic practices are inseparable from language processes. As Fairclough (drawing on Foucault) elaborates,

> Discourse as a political practice establishes, sustains and changes power relations, and the collective entities (classes, blocs, communities, groups) between which power relations obtain. Discourse as an ideological practice constitutes, naturalizes, sustains and changes significations of the world from diverse positions in power relations. (Fairclough 1992: 67)

The circumstances of political and economic change thus require that we pay attention to the co-determinations of text, discursive practice and social practice. For me, this means that we cannot avoid paying attention to the dimensions, technicalities and complexities of language organization – which, according to Fairclough (1992: 75ff.), are vocabulary, grammar, cohesion (sentences and paragraphing) and text (or narrative). It also means that we must be alert to discursive practice, the context in which language is produced, distributed, consumed and interpreted. And it means, at a different level, that language is imbued with ideology and power and all that these create and amend.

Language is produced dialogically

On the surface, then, dealing with the language issue seems to offer the researcher and political activist abundant opportunity for building powerful new knowledge. Yet, and this is where it gets difficult, it is never clear where language originates nor is it certain what it is referring to. So, at the same time that we have an argument compelling attention to language, we have an understanding that, as a carrier of meaning, language is an uncertain and contested social domain. Fairclough (1992, 1995) sees this as a creative tension. On the one hand, drawing on Bakhtin (and the Bakhtin circle), Fairclough points to the *dialogic* quality of language: that language is a mediated social practice and therefore all text is fluid and uncontained. On the other hand, we need an explanation for how text is stabilized as a resource for powerful and governing groups and here Fairclough looks to Gramsci's concept of hegemony, being that 'stabilized configuration of discursive practices' (1995: 2). Allowing these tensions of language production – a co-presence of dialogism and hegemony – to infuse social process offers much potential for the development of a post-structuralist politics.

Language needs purpose

Post-structuralism is a project which exposes the role of language and its use as a power device in the full spectrum of human existence. Certainly, the process of deconstruction is a powerful political tool for the scholar activist, though inevitably as after-the-event exposé. Yet an activist post-structuralism can also drive political action such as has been the experience of feminism, post-colonialism and queer studies. These political projects are distinguished by a self-conscious development and installation of discourse, alert to the conjoint tasks of re-moulding existing language and inventing new language, and devising strategies for their deployment.

The following discussion arises from sets of events that have involved public engagements with corporations or corporate activity, which have taught me the importance of a self-consciousness about the role of language in scholar activism, and the effectiveness of adopting a post-structuralist approach.

Interventions

There are many calls for geographers to get involved in public comment (see Garcia-Ramon 2004). Too often, though, and good intentions notwithstanding, public engagement is seen to follow a sequence from research, the development of academic text then the development of script for public release. The alternative, of course, is scholar activism where the research process, its texts and its engagements are co-determined; where purpose leads, and persuasion is crucial at every step (Gilmore 2005). Moreover, as Gregory (2005) demonstrates, research proceeds with awareness that the public and the public audience are framed by the research process such that assuming the pre-existence or independence of either is fanciful.

Like most of life's wisdoms, my understanding of this co-determinacy comes retrospectively. As a researcher, I have had a long-standing interest in corporations and the processes of industrial restructuring, involving, as you'd expect, the viciousness of capital's mobility, the vulnerability of labour and the awkwardness and ineptness of a schizophrenic state. The data collection phase of this research has involved, typically, the collation of information about a restructuring event by scouring, somewhat ironically, *publicly* accessible material from financial media, corporate documents and government inquiries. This has then been sorted into a storyline, rehearsed through face-to-face interviews, mostly with corporate managers and trade union leaders, and another tale in the industrial restructuring narrative of the progressive academy has emerged. Mostly, however, its academic readership is not substantial, apart from the more accessible pieces that find their way on to student reading lists.

As an academic in a regional university in Newcastle, Australia's stand-out twentieth-century industrial city, however, my tales commonly migrated to the public arena, as local journalists sought me out as a commentator on the succession of corporate rationalizations and closures that struck the Newcastle region during the liberalizations of the Australian economy in the 1980s and 1990s. Thus, I was enlisted as an expert commentator, though a naïve one. Two events exposed my naïvety. The first was an interview with a journalist which formed the basis of a newspaper article, the second the publication of a regional geography book. Both generated legal action where my academic skills of inquiry were scrutinized heavily. The incidents produced considerable self-examination of my standards and techniques in language use. Here I recount these two events, though in general terms since legal settlements in both instances require that details remain confidential.

The newspaper event

This event involved my acting as an expert commentator in a newspaper story about the closure of a factory. The factory was significant nationally, with a long history as a major employer of women in an industry sector which had received substantial government restructuring assistance. The factory was part of an intricate set of cross-ownership and supply relationships involving some of Australia's leading manufacturing entrepreneurs and retailers. I had tracked changes in this industry sector for at least a decade, received ongoing government research funding, delivered commissioned policy advice to governments and often gave background commentaries to journalists. The event recounted here occurred the day after the factory closed its doors following a protracted wind-down. The factory was the last of its type, superseded by off-shore supply chains.

The closure coincided with the announcement of a new package of substantial government assistance to the sector. The closure offered the opportunity, then, for me to reflect on the processes of restructuring which had occurred and make comments on the new assistance package. I remember being keen to argue that as a condition of being provided with financial assistance from the new scheme firms should be obliged to meet specific performance outcomes, especially in relation to employment security. I had rehearsed this view some months previously in a trade union submission to the federal government which relied heavily on commissioned work undertaken with some academic colleagues.

The newspaper chose to run the story on its front page. The story was also featured on those wire-framed posters that lean against newsstands and outside news agencies. The newspaper's main editorial that day strongly supported my argument for the need for specified employment outcomes. Notwithstanding the story's prominence, I saw my commentary as little more than providing a storyline *after* an event, and of drawing lessons for the future.

Two responses to the newspaper article therefore came as surprises. The first was a phone call to my desk from one of the major national players and a majority owner of the group of companies associated with the closure. I recognized his name immediately. This person's message and tone were scorching. He vigorously disputed the article's storyline and its supporting facts and claimed that the article carried the imputation of improper behaviour by the factory's owners and managers. I scribbled hurried notes as he talked and said little in response, being immediately alarmed by the prospect of litigation and shaken by the person's forceful, bullying tactic. The caller concluded by informing me a partner in the industrial group was actively seeking legal advice in respect of action against me, the university and the newspaper.

Late that day, I received a faxed letter courtesy of a top-tier law firm – as did the newspaper, an outlet of a national media company. In four detailed pages, the letter claimed the newspaper article and editorial from the morning's paper contained 'factually incorrect statements' and 'defamatory imputations'. The letter claimed that damage had been done to the company and the managing director, called for immediate actions to restrict this damage (such as by an immediate published retraction), and reserved all rights to institute legal actions for compensation.

Letters like these are arresting. Like this one, they are timed to arrive late on Friday afternoons leaving the recipient without access to legal advice and institutional assistance until the weekend is over. For me, the weekend was spent in self-examination of the storyline I had constructed, the veracity of its accompanying facts, and the imputations drawn, by me, the article and the editorial. My weekend also included frantic dredging and sifting of years of stored documents, files, publications and reports. The emotional impact doesn't need elaboration. The creation of anxiety was not accidental, the legal profession having a well-honed skill in crafting highly effective, fear-generating words.

My university, to its credit, acted promptly on the Monday morning, accepted that my actions as a public commentator were part of my employment duties and thereby took on the role as defendant to the claims. It took more than a year for the matter to be resolved, concluding with a 'Deed of Release' which records the conditions of resolution and the agreement of the parties to them, including that 'This Deed and all matters relating to this Deed are and shall remain strictly confidential'. It is a standard set of words used in these matters and limits my comments here.

The book event

In this case I had written a section of a contemporary regional geography book. Part of the contribution was the story of a local entrepreneur's pivotal role in transforming a regional firm into a significant national entity and then its hostile takeover by an international corporation. The entrepreneur's name had become iconic within the industry concerned, reflecting his bold takeover moves and unconventional new ventures in the heady days of 1980s capitalist experimentation. My contribution recounted the spectacular rise of his company, described the circumstances surrounding the foreign takeover, the loss of regional control and the inevitable employment rationalizations and facility closures which followed. The story was placed in the context of widespread collapses of many Australian firms over-exposed to asset-price collapses and interest rate rises in the late 1980s.

About three months after the book's release, letters were sent from the entrepreneur and his solicitors to me, my university and the publisher, a prestigious national house. Like the letters referred to in the newspaper incident above, these letters alleged my contribution to the book contained 'factually incorrect statements' and 'imputations which are highly defamatory', and demanded that immediate steps be taken to 'arrest the damage caused by this publication', including recall of the book by the publishers. The letters also advised the 'intention to commence proceedings for damages'.

Once again, thankfully, my university accepted responsibility for any liability arising from the publication of the material and negotiated a settlement of the claim. Critical to these supportive actions, again, was an acceptance by the university of my expertise in the subject field and an acknowledgement that my contribution to the published material was a reasonable part of my academic role. Once again, though, details of the settlement of the matter remain confidential under the matter's Deed of Release.

Writing a newspaper column

Insights into the role of language in the newspaper and book incidents have become clearer to me via my newspaper column writing experiences. For the past six years I have written over 125 columns for the *Newcastle Herald* and the *Central Coast Herald*, Fairfax's highest-circulating Australian regional newspapers with about 150,000 daily readers. I write an op-ed, or opposite the editorial, column with the broad remit of being the paper's urban and regional affairs commentator. As an economic geographer with a broad interest in the impacts of the operations of corporations on cities and regions, the scope of my interest means I am rarely stuck for a topic.

The op-ed column emerged as a regular newspaper feature in the 1920s. It has endured as a reassuring feature of the modern newspaper, always there in the same spot with predictable length. Its purpose, among the range of objectives in a newspaper, is the expression of opinion. It is a separate and distinguishable thinking and writing event. For an academic, it involves a different writing genre. There is an expectation the text will actually be read. It brings to the writing process the presence of an audience that seeks more than a sampling of news or facts, one that is interested in 'more of a performance than an expression of intellectual argument' (Sylvester 1998: xi). In this way, the columnist becomes a 'self-referential element in the text along with the subject matter' with 'opinions ... given the imprimatur of a personality with whom the reader becomes familiar and friendly ...'. Thus the columnist is called to write about 'the inside of public matters: not what is secret, but what is latent ...' (George F. Will 1981, cited in Sylvester 1998: xvi), to express 'vigorous, splenetic opinion' (Sylvester 1998: xiii) or, in the words of legendary American op-ed columnist Max Lerner: 'To be a general columnist is to exclude nothing human from one's perspective, and to treat the personal seriously and the serious personally' (Lerner 1945, in Sylvester 1998: xvii).

In the context of declining newspaper readership and falling public interest in politics and non-celebrity lives, there has been increasing emphasis on the op-ed columnist as a way for newspapers to strengthen their relationships with readers. In the USA, 81 per cent of Americans read a daily newspaper in 1964, while this has fallen to 54 per cent today, with only one quarter of those aged 18 to 34 years involved (Cornog 2005: 43). As commercial ventures, though, there is little point in newspapers blaming the public for its lack of interest in substantive issues. Wold (2003: 25) notes that: 'There is no point in exhorting the public to care. This is the only public we have and it will do what it wants.'

The truth is a little more complex if we pause to consider the questions of who and what is the public. Gregory (2005) explains how publics are created by the discourses that embrace them, an insight just as important to a newspaper proprietor as to an activist. It bears mention that at the same time as newspapers have fewer readers, the production and dissemination of newspaper text in digital form means that most newspaper text is disseminated worldwide from commercial sites such as Factiva (a Dow Jones & Reuters company) or Lexis-Nexis; or directly from newspapers' own websites (Said 2001). Columnists' texts, therefore, have carriage through space and time. Celebrity columnists, such as Paul Krugman, Arandhati Roy, Germaine Greer and Will Hutton – each with compelling stories to tell in simple, original language with an uncanny feel for the interests and experiences of their readers – enjoy readerships around the globe. The public embrace of these writers confirms what is true of any writing: readers will attach themselves to the writers of enlivened, purposive text be they global celebrity columnists or columnists with mere local or national reach

Columns, then, are powerful devices for public engagement. I offer two comments on their nature and importance. The *first* is about content. Columns demand original writing and thinking in different ways from other newspaper texts. Invariably they involve storytelling, incorporating imagination, characterization, even a dramatic plot (Bartkevicius n.d.). So the columnist has a textual freedom unavailable to the journalist, a privilege which flows from the intimate relationship between a column writer and the reader (see Wheatley 2001, on Charmian Clift, especially Chapter 22). Moreover, while journalists create and follow the news of the day, a columnist can write without too much concern for immediacy, or agendas or contexts defined by others, including editorial directions.

Secondly, a columnist builds a reflective relationship with his or her readership. When I began op-ed column writing I used to write about an issue of the day with no one in particular in mind. My prime motivation was – and I retain this to a large extent – a 'sense of mission' (Confessore 2002: 19). Three years on, I can't think about a column without being conscious of my reader, a person who arrives at my piece with an expectation of continuing a relationship. I now realize that I have re-positioned myself as a writer: from an academic in a knowledge-privileged position to that of a writer in a conversation, a dialogue, with an audience. I realize, now, this audience, with its broader mix of language and life experiences, is different from an academic readership. So, too, I realize I earn my authority from things like honesty and a lack of vested interest in my text rather than asserting it as the consequence of my institutional place. Moreover, I am conscious of the tentative position I hold, with the paper's readers having final power coming directly from the ease with which they can turn a page and not read me at all.

The relationship with the reader, then, involves conversational ploys: anticipating a common interest area, devising storylines, selecting and implementing narrative devices, language structures and vocabularies to suit the circumstances, and so on. Yet the column's text is never my own. As soon as awareness of the audience − from the features editor to the readership − descends, my thoughts, my arguments, my evidence, my words, my stories become negotiated. This builds intensity in the conversation, especially when the text of a column validates readers' experiences. Confessore (2002: 20) says the conversation is heightened when the columnist writes things 'before it is okay to write them', or when the conversation contradicts pre-existing narratives that have tenuous hold on the public's sense of the believable or on its opinions (see also Wheatley, 2001). Often this requires columnists to expose themselves in two ways, both more pronounced than in academic writing: by revealing the way personal feelings affect the argument; and by being open to the attacks of antagonists beyond the protective walls of academia.

Conclusion: language awareness and the scholar activist

As scholar activists, we intersect with public dialogue at many levels. We provide facts and evidence, develop arguments, initiate narrative structures to clarify and interpret events and experiences, and promote intervention strategies. Yet while we author and propagate these texts, we have little control over their interpretation, use or further dissemination. This creates risk. The knowledge created through the public dialogues that we help initiate is context dependent. An individual reader, for instance, might adjust his or her personal or ideological stance, while a corporate or government agent might consider a strategic, aggressive, response. The point is that the knowledge created, including the storyline through which it has passage, builds new meanings and performs new roles as it transfers across people and through institutions. Moreover, this involves inevitably a passage through different social and institutional domains that lack common, even similar, criteria for the verification of truth claims. Whether through a role as researcher, expert commentator or columnist, this places the scholar activist at risk as qualitative differences between intention and interpretation widen.

This means that a scholar activist's public claims and arguments will inevitably be challenged. Invariably, this is done using objective criteria: the accuracy of facts, the legitimacy and veracity of sources, the fairness of a narrative frame. The dilemma, and the risk, of course, is that a post-structuralist explanation of the knowledge creation processes involved is an unlikely defence within any tribunal established to examine truth claims. Here, George Orwell encourages ruthless self-consciousness:

> Every line of serious work that I have written since 1936 has been written, directly and indirectly, *against* totalitarianism and *for* democratic Socialism, as I understand it. It seems to me nonsense, in a period like our own, to think that one can avoid writing of such subjects. Everyone writes of them in one guise or another. It is simply a question of which side one takes and what approach one follows. And the more one is conscious of one's political bias, the more chance one has of acting politically without sacrificing one's aesthetic and intellectual integrity. (Orwell 2004/1946: 8, original emphasis)

The existence of an objectivist–relativist tension, outside our control, then, means scholar activism is always immersed in language tension. Addressing this tension, though, involves practical rather than intellectual or ideological actions. There are three considerations here. The first stems from understanding the reflexive nature of text construction and that there is a force of reputation that flows through text. An audience has a pre-existing propensity to believe a commentator's stories. For an academic, reputation is built from predictable things: exhaustive inquiry, attention to detailed records, strong narrative structures and powerful use of accessible language.

The second involves the development of trusting relationships with those who disseminate our work and these are, most commonly, people who work in the media and thereafter our readers and listeners. In my experience, the academy has an unfairly negative view of the work of editors and journalists, rarely understanding their work processes. The rules for effective, trusting relationships with journalists are simple: respect the work and time demands of their job; provide simple, clear, consistent messages; avoid dramatic off-the-record claims if they are to be followed by moderated, guarded interview statements; and acknowledge the need for verification of truth-claims even if they are not asked for at the time.

The third consideration concerns *legality*. Effective, alternative stories of corporate behaviours, for instance, inevitably attract responses and these may involve legal considerations. This is why you must be attentive to the evidence that underpins your interpretations and arguments. You must also be attentive to your responsibilities and protections: of compliance with ethics committees, of the nature and extent of your institution's responsibility for your public actions and of the expectation and desires of your audience, including your peers, and of the need to act officially, that is, by proper use of the formal roles and privileges that flow from your 'office'. Thereby, in the event that legal action is taken against you and your claims, you can be confident of a satisfactory resolution. Suffer the pain of a reconciliation process silently. Largely civil actions are an amoral, non-judgemental process and they spill little or no ideological blood.

By way of encouragement, though, creating public language teaches a lot about writing beyond the narrow genre that we confine ourselves to normally. As an academic engaged in wider public discourse, you will come to share your authority and acknowledge that your reader has a position, an intelligent position, an interest in ideas, a fascination for language, a desire for engagement. It will make you build lively ideas and arguments and stories, stripped bare of jargon and cliché. As economic geographers we have much to say, stuff of substance that the reading public is intensely interested in. As well, we have wonderful, instinctive, storytelling capacities, with stories that contain sophisticated messages that can start important conversations. Their power, though, depends on our use of accessible language; but not bland language, rather language that is rich in meaning and provocation. The task of a scholar activist, then, is to give intensely self-conscious attention to words and their use in a contingent world infinitely bound by narrative accounts of existence.

Acknowledgements

This chapter draws heavily on discussions with Pauline McGuirk, Bob Fagan and Elissa Sutherland. I thank the editors for their inspiration, guidance and patience.

10 THE PLACE OF PERSONAL POLITICS

Jane Wills

Over the past twenty years the shift to post-positivism across the social sciences has opened up space for new methods of social research. Once the orthodox model of research – based on a disembodied neutral observer collecting data from a world apart from themselves – had been rejected, scholars were able to make the case for alternative epistemologies and methods. We are now urged to reflect upon the nature of our engagement and position in the world, its influence on the data and knowledge produced, and its silences and partiality as well as its 'truth'. However, human geographers have been much better at addressing the internal politics of the research process than they have in exploring personal political issues, and in particular, the choices we make about what we study, how we do it and what it is for.

Our research is necessarily shaped by what Sayer (2005) has called 'lay normativity'. This involves the way in which our values and moral/ethical standards shape the way in which we look at the world, our selection of research topics, the conduct of our research and what we do with the findings. Yet despite the impact of post-positivism on both geography, and the politicization of the research process itself, surprisingly little has been said about our *motivations*, the way these affect what we do and the arguments that we might want to make. While it is now common to situate our research *subjects*, geographers have rarely turned the table on themselves: we have not located ourselves as researchers, highlighting the way in which we formulate and conduct research from our own very particular personal and political positions. Here, I reflect on my own experiences of doing and using research.

In doing so, I first explore the motivations for doing academic research within the context of calls that have been made for greater attention to the 'normative' (the *what should be*) in the discipline (Corbridge 1998; Sayer and Storper 1997). At least some of the recent disciplinary anxiety about the hazy connections between scholarship and activism and between research and public policy reflects the need to address the question of motivation. Given the collapse in left intellectualism in the academy, we need to be clearer about why we are doing what we do.

Second, I explore the ways that these issues have influenced my own research into trade union organization. Writing this chapter has forced me to think about my own motivations and interrogate why I believe that trade unions and the broader labour movement are important (and interestingly, how my views about the movement and its role have evolved in the fifteen years I have been working on them). In the process, I have also had to clarify what I am trying to do with the research, the relationships it has allowed me to create and the findings it has produced. In short, I have had to face up to my own 'lay normativity' and the personal politics of what I have been trying to do. At the end of this process, it strikes me as remarkable that as academics we are able to engage in the work of research and scholarship without having to acknowledge these questions, even to ourselves.

Thinking about why we do what we do

Over the past decade or so, there has been considerable collective anxiety within critical geography about our research motivations: exploring whether we should simultaneously be activists, working in and out of the classroom to change the world and the way people think (Blomley 1994, 1995; Castree 2000; Tickell 1995). Similarly, we have agonized about whether geographers can and should seek to influence the wider polity, and the formulation and implementation of public policy (Dorling and Shaw 2002; Martin 2001; Massey 2000, 2001; Peck 1999). Both sets of debates are focused on the role of the academy in studying, explaining and changing the world.

Such anxiety is perhaps not surprising given the structural conditions of academic life. In North America, young staff – arguably the most politically engaged – must publish according to clearly understood performance criteria to get tenure, while in much of the rest of the Anglophonic world, individuals and departments are routinely subject to external audit of 'quality'. In the UK, for example, the periodic Research Assessment Exercise judges the quality of research publications against a four-point scale, three of which are variations on internationally significant contributions. At a time when we need to publish in order to retain our jobs and make progress at work, and we all have material and psychological self-interests in publishing, it is harder than ever to step back and assess the motivations we have. While Doreen Massey (2001: 12) has recently argued that: 'I do believe that your next article, or project, should derive from some passion greater than simply adding another item to your CV or to the Departmental Output Count'. The motivation for doing what we do is likely to be a complex, partially sub-conscious, mixture of factors, including (among other things) the pursuit of collegiality, enjoyment, fashion, knowledge, politics, power, recognition, respect, security, self-fulfilment and status.

At least part of our motivation will relate to our own 'lay normativity' – the values and moral/ethical standards that help shape what we think important, judge right and wrong, and feel ought to be done (Sayer 2005). Over the past decade or so, a growing chorus of voices have been calling for increased awareness of ethics in our research and for a renewed focus on normative questions in the social sciences (Corbridge 1998; Sayer and Storper 1997). Although more applied areas of human geography (such as environmentalism, or development) have long been concerned with the 'positive' (description and explanation) and the normative, the core theories, concepts and approaches of the discipline have not. Since the radical turn of the early 1970s, economic geography has been much stronger in excavating the impact of structural and systemic processes and their attendant power relations on the human landscape than in positing ways out of the mess. While we may be happy to explore the 'laws' of uneven development, spatial divisions of labour, agglomeration, competitiveness, inequality and the struggle to produce the scalar architecture of our world, collectively we have less to say about *how* to alleviate the resultant socio–economic injustice.

I do not believe that this reflects a lack of 'lay normativity' in human geography. The strength of the arguments made in print are testament to anger, frustration and hope about the current state and future of the world. Such sentiments come from knowing that things could be different, and arise from personal moral/ethical judgements about what is wrong with the present state of affairs. Yet this 'lay normativity' is rarely articulated and rarely manifest in a normative focus and/or argument arising from such research. As Sayer and Storper (1997: 1) suggested nearly a decade ago:

> Any social science claiming to be critical must have a standpoint from which its critique is made, whether it is directed at popular illusions which support inequality and relations of domination or at the causes of avoidable suffering and frustration of needs. But it is strange that this critical social science largely neglects to acknowledge and justify these standpoints.

We might, therefore, conclude that theoreticians of human geography have been better at identifying and explaining the ills of the world, than at identifying the standpoint from which they are making such claims, or what might be done in the future. I think this is, in part, due to the fact that during much of the latter part of the twentieth century, left/critical intellectuals shared a broad set of perspectives that included an understanding of the importance of class in capitalist society, the role of the state, the significance of race, gender and sexuality, and the need for a socialist political party (albeit possible to adopt a social democratic or revolutionary model of change). Broadly speaking, the left (in and out of academia) was on fairly secure ground and there was little need to articulate the particular position from which people spoke. However, since the fall of the Berlin Wall and the renewed vigour of political-economic processes of neo-liberal globalization, both social democratic and revolutionary routes to socialism have been largely discredited. The old orthodoxies of the left, bound up in categories like capital, state, party and class have withered away and left intellectuals are now rather marooned (Benton 2004). As Sayer and Storper (1997: 2) put it: 'On the left, the years of complacency about alternatives and ethical positions are coming to an end.'

It is on this ground that Corbridge (1998) argues that Harvey's *Justice, Nature and the Politics of Difference* (1996) is unsatisfactory because it is premised on the fact that the 'end point' of socialism will dissolve all of the problems we face. For Corbridge, the book fails to undertake the hard work of engaging in practical ideas about the reformation of capitalism and the need to challenge the pervasive ideas of the right. In this context, there is now a slow process of reformulation underway in which left intellectuals (including many of the leading thinkers in human geography) have started to rethink the ground from which interventions are made. Thus far, a number of responses have become evident in human geography, including a renewed interest in utopian thinking (Harvey 2000; Pinder 2002, 2005); the revival of interest in moral geographies (Proctor and Smith 1999; Smith, D. 2000); the use of post-structuralist tools in an attempt to validate the local and the micro as sites of political intervention (Gibson-Graham 1996); the use of religious principles in research and practice (Cloke 2002); a move towards political pragmatism and policy engagement (Martin 2001); and rethinking the political concepts and strategies of the past (Massey 2005).

Although there is a widespread resistance to normativity in the academy among Marxists (who often equate moral reasoning with powerful bourgeois interests and who have entrenched – if somewhat rhetorical – beliefs in the 'end goal' of socialism), and post-structuralists (who (usually) hold that normative reasoning will necessarily be particular and reflective of power relations), Sayer and Storper (1997) argue that careful normative thinking is a way forward for academic research. I fully concur with this view because whatever the risks (and history teaches that there are many) there are greater risks in failing to act. Yet we need to know more about the implications of undertaking more normatively-sensitive and focused research. Is it necessary to acknowledge who we are and our political positions when doing this kind of research? Are we able fully to identify our motivations for doing research even if we wanted to? How is our research shaped by our values and moral/ethical judgements, and how do we know? How political can we be about what we say and do with research? And, ultimately, would this approach actually make any difference to the practice and outcomes of research? Sadly, as I've written this chapter, I have come to realize that I have very few answers and many more questions. Here I hope to contribute to further debate and in the meantime, I reflect – in a very low-key way – on some of these issues through my own research and experience.

Research into trade union futures in the UK

Preamble

In March 1998 I started work on a proposal for a three-year research programme on trade union futures in the UK, for submission to the Economic and Social Research Council (the main funding body for academic social sciences in the UK).[1] I intended to explore the two main options for British unions at the time – partnership and organizing – in an atmosphere of renewed optimism associated with the election of the Labour government in May 1997. While partnership involved unions and managers working around an agenda of 'mutual gains', organizing was focused on workers' self organization and action around collective concerns. I planned to illuminate the spatiality of these two strategies through in depth empirical research: to look at how each strategy varied in its impact and implications across space; how each approach was affected by the particularities of place; how the two strategies were complementary and contradictory; and how these factors would, in turn, reshape national policy and debate. The project was formulated as a 'geography of' trade union renewal in the UK, framed in terms of previously published work of my own and others working in industrial relations, sociology and human geography and it highlighted the scope for disseminating the findings to trade unions, government bodies and non-academic organizations. In a very real sense, my aim was to extend academic knowledge and debate, in and beyond human geography, while also contributing to the policy development of the trade union movement itself.

Personally, I was excited by the move to organizing. The British trade unions had been managing decline since the early 1980s, responding to each political and economic assault as it came. The challenges of the privatization and 'marketization' of public services; the wholesale loss of manufacturing capacity and the concurrent explosion of employment in private services; intense global competition; the changing legal environment and numerous industrial defeats had left little time or energy with which to develop a strategy to stem and reverse the decline in trade union membership and political influence. The vague hope seemed to be that the eventual election of a Labour government would allow the unions to emerge from the bunker and maybe then, they could more confidently address the question of growth. However, by the mid–1990s, after three Conservative governments and in the face of barely disguised hostility on the part of *New* Labour (the moniker adopted by Tony Blair as a means of distancing his leadership from Labour governments of the past, the trade union movement and left–wing elements of his party), a group of far–sighted union leaders began to proselytize the need for renewed attention to organizing, citing similar attempts underway in the USA and Australia. Eventually, what became the New Unionism project under John Monk's leadership of the Trades Union Congress (TUC) was established to:

- Promote organizing as the top priority and shift the unions towards an organizing culture.
- Increase investment in recruitment and organizing, strengthen lay organization and promote the use of dedicated organizers.
- Strengthen existing bases and break into new jobs and industries.
- Sharpen the appeal of trade unions to 'new' workers, including women, youth and those at the rough end of the labour market. (See Heery 1998; Wills 2005)

Although never formally articulated at the time, looking back, the research I planned to do was partly driven by my excitement at these developments and genuine curiosity to see what happened. Moreover, although I was most interested in the organizing agenda, I wasn't convinced by those on the left who simply rejected partnership as class collaboration, not least because all collective bargaining is founded on having a relationship with managers and the outcomes of all such relationships are shaped by the balance of power relations. I was also keen to use the research in order to assess genuinely what worked in practice – and the strengths and weaknesses of each approach – and feed this back to the trade unions to shape policy as it evolved. Finally, I was motivated by the wider political implications that a revitalized trade union movement would have for the UK. If workers were better organized, labour interests – broadly defined – could be more clearly articulated and defended in the national polity. As a mass membership organization with its own resources, the trade union movement could shape the political climate of the UK through its influence with government, employers and the wider public (as it had done before). Even now, the trade unions represent some 7 million workers and have assets and human resources worth millions of pounds. The future of the movement mattered to me and reflected my own values and moral/ethical standards concerning questions of justice, equality and mutual respect.

In the spirit of openness that I called for at the start of this chapter, this political perspective came from a mixture of involvement in Marxist politics from my late teens to mid-twenties, preceded by my upbringing in a Methodist home. Getting involved in socialist politics – after first entering the peace and women's movements – and then getting caught up in the miners' strike (1984–5) meant that it was just about plausible to join an organization that claimed the organized working class as the agent of the socialist dawn. The miners' strike dominated the news for at least a year of my time as an undergraduate. While I was avidly reading radical geography and Marxism, I watched the class war on television. The country was divided and, then at least, class did seem primary as a source of identity and political agency. Moreover, my few years as a political activist led me to get embroiled in the day-to-day politics of trade unionism and I learnt that more modest gains, such as improved working conditions, pay and benefits, or challenges to racism and sexism, were possible through collective organization at work.

Writing my research proposal more than a decade later, I had long-since recognized that the organized working class would not lead us to the socialist dawn, but retained a strong political attachment to the trade union movement. Part of my motivation for undertaking the project itself was probably to try to come to terms with questions of class and political agency. Deep down, I knew that I needed to clarify my own thoughts about the trade union movement, its role and its future in the UK.

Doing and using the research

Once underway from October 1999 it became clear that the research problematic was far more complex than I allowed for in my research proposal. The developing research programme depended strongly on negotiating access to key sources and developing research relationships, the art of the possible and a certain degree of good luck. In order to research the twin models of organizing and partnership, I needed access to the trade unions from the top down, to talk to those involved in the development of the New Unionism and partnership projects at the TUC and, in particular, I also needed access to those unions and leaders who were experimenting with the new approaches. Thus, I had to find people who were willing to enter into a research relationship, and from there, identify examples that were particularly worthy of further research. By definition, I had to contact trade unions known to be experimenting with partnership and organizing – and there weren't many of those at the time – and then find people who were open and willing to being involved in the research.

Furthermore, even if I was able to forge positive links, these relationships didn't always bear fruit. More than once, I started research work, or undertook a number of interviews, and then hit a dead end. This was partly because union personnel might move and their replacements were less favourable to doing research, or because particular organizing campaigns fizzled out on the ground, or because they stretched beyond the life-time of the research and into my subsequent maternity leave. In one instance, I completed some interviews on the geography of organizing activities with key unions for a document commissioned by one department at the TUC, only to find that those higher up in the organization chose not to publish my submitted report.

In the end, however, I developed relationships with a number of trade unionists in the UK, some of which lasted longer than others, and a number of which continue today.[2] The research developed its own momentum, based on these relationships and the activity of the organization, the balance between their needs and my interests, and what was possible in practice. Over time, I gained a better understanding of the strategies being deployed, their strengths and limitations, and greater knowledge about organizing workers in other parts of the world. Ongoing research in the USA, in particular, highlighted the need for unions to develop a new geographical imagination to reach low paid service workers such as janitors or hospitality workers. These researchers and activists argued that organizing at the workplace was no longer enough to win trade union organizing campaigns among these groups of workers, and my own experience of the limits of organizing campaigns among hotel and travel trade workers in Britain reinforced this for me (Wills 2005).

Despite this literature, and the impact of experience, the British unions have not grasped the full scale of this geographical challenge. In the main, the unions have focused on the workplace as the appropriate scale to launch and manage campaigns – in part, reinforced by the new legal entitlements to recognition introduced in the Employment Relations Act 1999. The unions have also largely failed to appreciate the way in which widening the scale of any organizing campaign can facilitate the identification of new allies (such as community and faith institutions, students and others interested in social justice) to add weight to their cause. Through the process of research, reading, thinking and talking, I began to realize that more should be done, and sought to develop a broader understanding of the geographical implications of organizing in the UK. Rather than simply doing a 'geography of' research project, as envisaged at the outset, I began to explore the geographical imaginations, structures and strategies of the trade unions in the UK.

By drawing on additional research with an international trade union body, looking at organizing in the hotel sector, and completing some research with the East London Communities Organization's (TELCO's) Living Wage campaign, I began to develop an argument about the need for unions to build networks with each other and their allies as part of an *extra* workplace strategy for renewal. Seeing evidence of gains as a result of rescaling union practices, I was able to make a stronger argument about geography than originally anticipated. Echoing debates about scale that were exercising human geographers at the time, it was possible to argue that trade unions could not expect to organize workers without expanding the geographical ambition of their efforts. Rather than using traditional approaches to organize in the workplace, the private services sector in particular demands a labour-market approach. Workers move regularly, many work for subcontractors and it is easy for employers to resist local efforts to organize their workers. Moreover, in a market-dominated economy, the increased wages won by union members can erode the market share of their employers, making it necessary to organize across the whole labour market in sectors such as contract cleaning, catering, caring and hospitality.

Rather than simply cheerleading those advocating organizing, as might have been expected, I found myself able and willing to make a political argument about the need for unions to rethink their geographical imaginations, structures and strategies in order to reverse their decline. I published a number of non-academic articles in *Red Pepper* and *Unions Today*, and wrote a longer pamphlet for the Fabian Society, making this case (Wills 2002). In addition to undertaking an academic project, I was able to make a public argument about what I felt the unions needed to do. The pamphlet was launched at the TUC Annual Conference in Blackpool in 2002, and it has been picked up by a few individuals in different trade unions. It has also led to invitations to speak about community–union relationships to different groups of trade unionists. As academics, we have the time and resources to think and research issues in depth. If we then have something useful to say, we should say it. Indeed, it seems rather strange that so few academics publish in order to communicate beyond their own field.

Conclusion

This story might be helpful in thinking through the questions with which I began this chapter. On reflection, I have realized that I was engaged in normative – or action – research (and for more on this, see Wills with Hurley 2005). My 'lay normativity' prompted me to explore trade union futures in the first place, the research was then driven by my own need for answers about strategy and practice, and I used the findings to make a political (normative) argument about what the unions needed to do. Significantly, I finished the project with less political confidence in the organized working class than when I began. The scale of the problems faced and the weakness of the tools being used to respond made me question the very future of what we call trade union organization. The research also blew away residual ideological cobwebs of my own about trade unions, collectivity and political agency. Writing this chapter has clarified this further, and it strikes me as odd that I have never been challenged to do this before.

My realization about the scale of the challenges and the weakness of the trade unions in Britain helped to fuel my research and my use of the results. It drove me to try to articulate what I thought ought to be done. I did this on the basis of genuinely fresh thinking on my part: I had not found what I expected to find, and I had identified a much stronger argument for geography than I would have thought possible. Too often, we are constrained by the belief that political engagement compromises our academic detachment and rigour. Thus, even though the post-positivist environment highlights the politics of the research encounter, the dilemmas of representation and the partiality of our accounts, we rarely talk about our own political motives and passions in doing research. This helps to explain the rather anguished debates about scholarship and activism, research and policy. There is a thirst for politics and passion about what we do. However, it is also significant that in all the debates about scholarship and activism, research and policy, contributors have focused on the way in which we might work with others as activists/action researchers, use research to help others, feed ideas into policy development, and/or mobilize ideas in the classroom rather than make our own arguments. With the luxury of three years of research on trade union organization in the UK, I was able to stand back, reflect on my findings, reading, thinking and talking, and make a political argument that the movement needed to take its geography more seriously. The short pamphlet I wrote for the Fabian Society is out in the public domain, and, even though it is rather thin and not widely read, it provides a way to reach those in a position to act.

NOTES

1 Geographies of Organized Labour: The Reinvention of Trade Unionism in Millennial Britain (ESRC R000271020).

2 Some of these relationships have been sustained through research studentships and include work with the regional TUC (Holgate 2004), the Citizen's Organizing Foundation (Jamoul 2006) Paula Hamilton and the International Transport Workers' Federation.

11 LOCATING THE THAI STATE

Jim Glassman

The nationalist modifiers that habitually prefix the word 'state' – in my case, 'Thai' – impart a considerable geographic specificity. The Thai state seems to be delimited to the space of Thailand. Yet, without rejecting the salience of the various political projects called 'nationalist', I have come in the course of my research to doubt that there is any such singular, geographically bounded Thai state. I reflect on this process of questioning in three interconnected ways here, each speaking to the politics and practice of research. First, I identify distinct ways in which 'the Thai state' can be unbundled as an object of analysis and the associated methodological issues that have come up. Second, I note how even unbundling the Thai state forced me to confront the residual solidity of a certain type of 'national' state. Third, I examine some of the politics surrounding such unbundling, including my unpaid political debts to those whose struggles I have studied.

The politics and practice of researching the Thai state have been, for me, deeply situated, and as such I speak here in a very definite first person. The research challenges I have confronted have evolved in response to specific developments that I could not foresee, making them deeply situated because of both my positionality and the constant, ongoing transformations in the 'object' of my research. I seek to capture some of this situatedness and fluidity, starting with myself.

Positioning oneself

Gillian Rose (1997) has aptly warned of the conceit that the motivations shaping our work can be simply identified and declared. Yet recognizing that our motivations are not completely self-evident should not absolve critical scholars from attempting to note how who they (think they) are influences their knowledge production. While much that has animated my own research may escape my comprehension, I can identify two factors that I consider important in situating this research – and shaping both the object of analysis and the theoretical orientation I have carried into my work.

Before deciding to study Thailand, I spent much time studying Central America – as a political activist opposed to US imperialism in that region.

My focus was on Central America as an object of US foreign policy, and the disastrous effects of this policy. Coming into the Central America movement in a milieu in which many had graduated from anti-Vietnam War activism, I attempted to understand the present (US policies in Central America) through the lens of the immediate past (US policies in Southeast Asia). I thus read much on Vietnam, Cambodia, and Laos. This grounding of my interest in Southeast Asia in opposition to US policies has exerted an ongoing influence on my approach to the Thai state.

The decision to study Thailand, when I went to graduate school in the 1990s, was driven by personal filiations. My partner grew up in Thailand as part of the Thai student movement during the 1970s that opposed US militarism in Southeast Asia and the US-backed Thai military dictatorship. Notwithstanding noteworthy successes, the movement was violently repressed, and in Thailand there still are many from this generation who feel a sense of having yet to fulfil the projects they started decades ago. A considerable number are today in positions of power, wealth, and influence – by no means all still committed to social justice – and they have provided me ready, interesting, and diverse points of access into contemporary Thai social struggles. Several – notably, internationally renowned contemporary public intellectuals such as Thongchai Winichakul and Kasian Tejapira – have also provided me with ready and stimulating perspectives on the state and social struggles in Thailand (Thongchai 1994; Tejapira 2001).

These aspects of my positioning have interacted to shape my research agendas. While much of Latin America was suffering a 'lost decade' during the 1980s, the US-backed Cold War states in Southeast Asia were undergoing rapid economic growth and industrial transformation, leading the World Bank to declare them part of the 'East Asian miracle' (World Bank 1993). Thailand had the most rapidly growing economy in the world between 1986 and 1996, and as a Cold War state that had hewed especially closely to the preferred US economic development model, it made for an interesting case study. I consciously made a decision to change my analytical focus in two ways: from disaster stories of US 'intervention' to a success story, to see what it was that was being constructed as the best advertisement for following the preferred US development agenda; and from studying a country from the outside in (i.e., from the perspective of effects of US foreign policy) to studying it to some extent from the inside out (from the perspective of how various 'local' actors engage the processes that have been called development). I did not abandon the perspectives that originally situated my interest in development issues. I studied a success story with a sense that much of what was called development was for many people a disaster, and while approaching it from the inside out, I retained a sense that 'external' forces were crucial in shaping 'local' actions.

These two factors found their way directly into my theorization of the issues, and are thus complicit in my own production of knowledge about the Thai state. I do not find this a weakness: situatedness and partiality is inevitable, and being forced from the outset to recognize an 'orientation' to my research has lent greater clarity to my understanding of the research process and its limits. Moreover, starting out with an 'orientation' has not prevented me from feeling the need to revise what I thought I knew to be the case.

Unbundling the Thai State

I went to Thailand committed to using something akin to Michael Burawoy's extended case method (Burawoy 1991, Burawoy et al. 2000). Here, a case study should generate empirical evidence that produces modifications and/or refinements in the theoretical perspective employed while conducting the empirical research (see Yin 2003). Burawoy's approach can be employed using any methods that generate empirical evidence. I emphasized key informant interviews (e.g., with government officials, factory managers, labour union representatives), basic data collection and analysis (e.g., on wages, investment, profit rates), and considerable contextual, background research (e.g., archival work, geographical-historical reading and analysis).

I went to 'the field' armed with this approach, and long on theoretical perspective, having brought along a fair amount of social theory from a previous degree in social and political philosophy. I was quite short on empirical grounding, and found the extended case method to be especially useful because expectations that informed my earlier theorizations turned out to be based on limited understanding of realities 'on the ground'. Empirical work thus forced constant reconstruction of the theoretical arguments that frame my research. This is clearly a process with no (theoretical) end in sight. Let me note just a few of the (ongoing) twists and turns.

After doing a preliminary field research stint in Thailand, I co-authored an article with Abdi Samatar on differing theorizations of the state in development geography (Glassman and Samatar 1997). We argued that a series of factors seem important in determining the degree to which states are successful in fomenting industrial transformation and sustained economic growth. We tried to bring concepts we found valuable in the neo-Weberian literature on East Asian states (e.g., Wade 1990) into contact with ideas from Marxist and neo-Marxist state theories and studies of other regions. As part of the conclusion, I applied some of these concepts in a brief summary of development issues in Thailand, noting how, as of late 1996, the Thai economy was faltering and seemed to be unique within Southeast Asia for the degree to which it was being abandoned by international financial investors, perhaps indicating something uniquely problematic about the Thai state. No sooner had I concluded my field research in Thailand during 1997, however, than the Thai baht was unpegged from the US dollar and plunged rapidly, leading the entire region into a significant economic crisis. Clearly the problems that various investors thought they saw in Thailand were *not* unique after all!

Some of my research had already been leading me to question how I was applying the concepts I had put forward for understanding the development performance of states. I had begun to feel, from experiencing the evolving economic crisis in its regional and international dimensions and from talking with people inside and outside Thailand, that my sense of the Thai state was too *nationalist*. It seemed less fixed, less coherent, and less *Thai* than I had imagined. In the various bureaucracies responsible for development planning, I encountered official after official trained in Western universities, brandishing neo-classical economic theories, rational choice paradigms, and the like, and speaking the language of development honed in international development institutions like the International Monetary Fund (IMF), the World Bank, and the Asian Development Bank (ADB).

What was uniquely Thai about this state other than its formally designated space of authority? What was I to see national development as entailing if a planner could tell me that the growing income gap expected from a particular project should be acceptable to those who would be relatively disadvantaged as long as their absolute incomes increased? Or if a planner could tell me that a particular industrial strategy was expected to increase income disparities but was favoured because it brought rapid growth rates? Or if a factory manager for a foreign firm in one of the favoured industries could tell me that the company located in a state-promoted area of Northern Thailand to avoid Bangkok labour unions, and employs young women whom they thought would be easier to control? Or if an occupational health specialist could detail for me the ways in which these same rapidly growing industries produced numerous broken bodies, even as workers and health specialists within the state struggled against other state agencies to regulate industrial growth in ways that would limit this human toll?

Thai state agencies seemed staffed with allies of internationalized capital and had become adroit at speaking the neo-liberal language of 'trickle down': 'some have to get rich first', as Deng Xiaoping put it. 'Thailand' had been highly successful at this strategy, but was this 'national' development, or merely capitalist development within particular sites of an already internationalized political economy? I decided the latter, and that the Thai state needed to be studied as an internationalized entity. I thus came to write about the 'internationalization of the Thai state' (see Glassman 1999), borrowing from Nicos Poulantzas (1975). And, thus, a happy Burawoyan turn: I had revisited and revised slightly earlier conceptualizations of the state, based in part on fieldwork and new empirical information. But nothing is ever entirely happy and there probably are no endings.

I felt that my key informant interviews, data collection, and historical-structural analyses had enabled me to assemble a strong case for this internationalization. I traced the historical processes by which key Thai development agencies had been formed in collaboration with international actors – including the US state during the Cold War, showing how the major players had received 'Western' training and had implemented a development trajectory underpinned by international capitalist agendas. Thai investors openly courted and benefited from joint ventures with Japanese, US, Taiwanese and numerous other foreign investors, aided and abetted by the Thai development bureaucracy. Further, the entire agenda frequently was enacted at the expense of various social groups in Thailand (especially peasants and industrial workers).

Yet in fundamental respects my research 'sample', and my very positionality as a researcher in Thailand, inclined me towards this sort of result. I had sought out the development agencies that happened to be the most internationalized. As a *farang* (a Westerner) and a white male, I was directed quickly and readily towards those people who could communicate best with someone like me, and at the head of many agencies was frequently a bureaucrat with extensive international experience, fluent in English, well schooled in neoclassical economics or Western business management techniques. Naturally, I found evidence of internationalization. But how reflective was this of an entire, complex, bureaucracy?

Stumbling on 'the Thai state'

My connections with former Thai activists helped me to see things with a little more nuance. None doubted the international forces at work, but all had grappled regularly with agencies of the state that had much more uniquely 'Thai' attributes, including the monarchy. I was forced to recognize the Thai state as hybrid, internationalized but retaining strongly national – even national*ist* – dimensions.

This was powerfully driven home by my interactions with groups from Thailand's federation of activist organizations, the Assembly of the Poor (AOP). The AOP includes the Council of Work and Environment Related Patients' Network of Thailand (WEPT), which I studied, but also a large number of groups working on rural development and natural resource issues, such as opposition to hydro-electric dams that destroy fishing livelihoods.

Former student activist, Wanida Tantiwittayapitak, a leading spokesperson for the AOP, acquainted me with a number of AOP struggles, including, eventually, struggles against repression by a very 'nationalist' Thai state.

At the close of my field research in 1997, the AOP staged a 99-day sit-in at Government House in Bangkok, with tens of thousands of people demanding agreements on issues ranging from crop prices, access to forests, and dam decommissioning to compensation for injured workers, revamping of the occupational health and safety bureaucracy, and support for slum communities. This demonstration focused – with tactical sophistication and some short-term success – on the Thai state. As was to become even clearer to me through subsequent interactions with the AOP, internationalization did not mean that Thai state activities were mere reflections of the wishes of internationalized capital. As AOP members understood, the Thai state had to be dealt with in terms of its national and local specificities, including those aspects of its structure and behaviour that are idiosyncratic and relatively independent of internationalization.

Comprehension of this was forced upon me by how AOP struggles played out during the economic crisis and its aftermath. The AOP had negotiated a number of agreements with the government of Chaovalit Younchaiyudh in 1997, but when the crisis hit in July his government was replaced by the neo-liberal Chuan Leekpai government, which implemented an IMF structural adjustment programme and refused to honour the promises of the previous government to the AOP. Chuan's government began threatening violence against AOP demonstrators, as discussed below. These threats were fundamentally homegrown: no orders from Washington, DC, from international capital or the World Bank.

The unpopularity of the structural adjustment programme triggered a backlash, and Chuan's government was replaced in 2001 with a self-avowedly nationalist government headed by Thaksin Shinawatra. This further suggested residual 'national' power and autonomy that had to be accounted for within the internationalization argument I was presenting (see Glassman 2004). My theorization of an internationalized Thai state had to retain space for considerable national specificity and autonomy.

Political debts

Most of my research has been positioned in relation to former Thai activists in such a way that at least some groups in Thailand struggling for social justice would likely regard me as a potential ally. Moreover, public intellectuals and academics, whom I met independently of these former student activists, have frequently been people with strong commitments to these contemporary struggles. My positioning as a partisan scholar has also been complex, however, given that Thai state officials who don't know of these connections are more likely to see me as a conventionally sympathetic white, male academic. Rather than correct for this positioning, I have tried to work through it in various ways, including taking on the specific research challenges that arise when contemporary events of salience to groups like the AOP come to the fore. One cannot do research in this way, however, without incurring large political debts, which I feel unable to repay.

The first was incurred to workers in WEPT, and to members of the Arom Pongpangan Foundation, a Bangkok-based labour non-governmental organization (NGO). Given my position as a white, American, heterosexual male academic with relative class privilege, doing research on labour processes that disproportionately involve young Thai women working in export factories, I approached research on these workers largely through the mediation of labour activists in Thailand already involved with such workers. Professor Voravidh Charoenloet at Chulalongkorn University, President of the Arom Pongpangan Foundation, connected me with many activists and brought me directly into contexts such as meetings on occupational health issues involving WEPT members and other workers from around Asia (see Charoenloet 1989, 1998). Through Wanida, I met the head of WEPT, Somboon Srikamdokkae. Somboon, a laid-off textile worker who suffers from byssinosis, astutely and appropriately asked me what I thought the results and impacts of my research would be. With embarrassment, I acknowledged that I could not see exactly what these would be beyond academic publication. Such confessions did not deter members of the Thai labour movement and the AOP from sharing information with me, but I have found it continuously humbling.

Upon returning to the USA, I sought to overcome this personal sense of disempowerment. I talked with people from various US labour organizations and NGOs, including the American Federation of Labor and Congress of Industrial Organizations (AFL-CIO). A number of international labour projects were underway, but few had any potential salience for workers in Thailand. The reality I began to confront, quite simply, was that with the preponderance of foreign investment in Thailand coming from Asia, with US labour unions more concerned about competition from China, and with a weaker labour union movement in Thailand than in places like South Korea, little if any labour solidarity activity in the USA was (or could be) directly connected to labour struggles in Thailand. The one active US labour organization, the AFL-CIO's American Center for International Labor Solidarity, with a long history of support for major US foreign policy initiatives, is viewed with suspicion by progressive Thai labour activists (Glassman 2004). Thus the limits of my political solidarity were not merely a personalistic lament but fairly strongly grounded in a geographical-historical reality of the labour movements in the countries my research bridged.

   This political limitation turned out to be less marked in subsequent research on the Thai state, but it also has had politically humbling moments. During my dissertation research, I was affiliated with Chiang Mai University's Social Research Institute, headed by Professor Chayan Vaddhanaphutti. Chayan's research and activism, especially as head of the Ethnic Studies Network and as an adviser to the Southeast Asia Rivers Network (SEARIN), has focused on civil society groups in rural Thailand (Vaddhanaphutti 1984), whereas my doctoral research was on urban-industrial labour processes. Consequently, our conversations at that time did not lead to any collaboration, despite shared interest in Gramscian approaches.

   During 2000, Chayan was a visiting scholar at the University of Washington. We both followed with interest and concern the activities of the AOP, and especially the activities of anti-dam activists at the Pak Moon dam in Ubon Ratchathani province in Northeast Thailand – the specific group from which the AOP had been born in 1995. Chayan and members of SEARIN had been working directly in support of these anti-dam activists, for whom Wanida was a spokesperson, attempting to help them advance in a broader national and international arena their claim that this World Bank-funded dam had ruined their livelihoods and should be decommissioned (see World Commission on Dams 2000). At a particularly intense moment of the struggle, SEARIN workers began e-mailing Chayan that the protest village in Ubon Ratchathani was being threatened with violence by people working for the Energy Generating Authority of Thailand (EGAT). Chayan and I responded by drafting a letter that we quickly circulated to international Thai scholars, demanding that the Thai state not attack the anti-dam protestors and listen seriously to their claims. We did not expect the Chuan government to respond, but hoped that raising the issue in ways that implied international concern over the Thai state's response would potentially place the struggle on a slightly different footing. We submitted the letter to both the Chuan government and the leading English-language daily newspapers in Bangkok. SEARIN and other Thai organizations sent similar letters and to the Thai-language dailies.

   At the time the Pak Moon struggle heated up in 2000, I was already aware of it (Glassman 1996) and was especially interested in the alliances being formed within the AOP between anti-dam activists and groups such as WEPT. I began to study in somewhat greater detail political strategies of the AOP and the responses of the Thai state, leading me to spend time analyzing the 'scale politics' of the Pak Moon struggle and of Thai state agencies opposing it (Glassman 2002). It became apparent that the agencies of the Thai state with the most power to confront Pak Moon protesters are highly internationalized agencies, albeit with their own 'national' purposes. They draw on their international connections wherever they can, particularly for funding and political support. The AOP also uses international connections when possible, but typically has fewer resources to draw on. The Thaksin government played a directly geographical strategy against the AOP, hinging on implicit notions of what counts as properly 'Thai'. Thaksin called on international funders of Thai NGOs to withdraw funding for organizations supporting the Pak Moon villagers, while launching investigations into this funding. The hypocrisy of a government, which had drawn down US\$17 billion from the International Monetary Fund to bail out big investors, criticizing Thai NGOs for receiving 'foreign' funding was palpable, but only in so far as the question of the 'Thai-ness' of the Thai state could be raised.

The Pak Moon struggle thus served as a path into further research and reflection on both the internationalization and 'nationalism' of the Thai state, and my relationship with people like Chayan and Wanida – grounded originally in this political solidarity – allowed me to further my project of deploying the extended case method. But what have I been able to return to these activists and public intellectuals in exchange? Some villagers I met from Pak Moon in 2002 were aware that letters had been written on their behalf by academics from inside and outside Thailand, and were gracious in their statements of appreciation. Yet their efforts were dealt a severe blow by the Thaksin government during 2003–4 and there seems little that can be done at present to counteract this. Indeed, by 2005 the entire network of Thai activist organizations was facing a crisis, embroiled in an intensive process of self-criticism and strategic reappraisal (Rojanaphruk 2005). I am haunted that, notwithstanding whatever efforts I made, the Pak Moon struggle served to help me advance my career, even as the events I have studied destroy the livelihoods of those with whom I have tried to exercise solidarity.

Conclusion

I do not mean to impart undue pessimism, let alone perform guilt that I don't in truth feel. I feel that the research I have carried out provides a lens for viewing the Thai state that is of potential utility for both scholars and activists. I went to Thailand an outsider, drawn in part by the prospect of studying a state that had been advertised as a successful case of participation in globalization, but attentive to the real problems that people engaging globalization from the bottom of social hierarchies are likely to experience. I duly found evidence of both a highly internationalized state and of serious problems this state had helped generate for many people in Thailand. It might be easy to conclude that I simply found what I was looking for. But as I have tried to indicate here, I do believe that I was open enough, as Susan Hanson (2004) insists we should be, to actually find things I was *not* seeking. In any event, since I conceive this process of research as open-ended, I do not purport to have developed any permanent answers to particular research questions, but merely to have developed more detailed arguments. My prior politicization and research orientation has no doubt helped shape these arguments; but given what I see as the objectives and methods of the research, this does not strike me as problematic and has in fact been enabling of the arguments and evidence I have produced.

The politics are far more vexing, yet the sense of 'failure' should be leavened with realistic appraisal. I did not enter into these activities promising anything to those with whom I have attempted to exercise solidarity, and I do not think that most of their members would ever have expected much to come of my activities or those of other foreign scholars. If I have been able to do little to exercise effective solidarity, neither do I think my research has had very much in the way of negative effects. Indeed, perhaps the most chastening experience has been feeling my relative weightlessness within the social struggles I study. This has been important in understanding what could be realistically construed as the politics of my research. Those politics are certainly not nil, but neither are they what I might hope for. Yet they have made me part of an internationalized network that studies and critiques the Thai and US states – perhaps a practical and appropriately humble way to participate in efforts to construct something better.

12 POST-SOCIALISM AND THE POLITICS OF KNOWLEDGE PRODUCTION

John Pickles and Adrian Smith

For over forty years, communism, anti-communism, and the Cold War constituted the primary horizons within which social and economic life unfolded in the Soviet Union, Central and Eastern Europe (CEE), and beyond. If the end of the nineteenth century had seen the arrival of industrial capital and a working class organized around socialist principles, the end of the twentieth century saw this geopolitical struggle undermine both the economic and ideological power of a state-led industrial socialist project. The collapse (like the advent) of soviet-style socialism was thus an event that transformed the political and economic landscapes of Europe, but one that also has had wider global ideological and theoretical effects. Not least, these were felt through the global ascendancy of neo-liberalism.

Central to our various research projects on the geo-economies of 'post-socialism' has been an attempt to negotiate four sets of influences. The first is the combined and uneven consequences of projects of shock therapy, structural adjustment, privatization, and European Union (EU) integration. The second relates to the value of grounded fieldwork that is attentive to locally specific cultural practices and the politics of knowledge production. The third focuses on the intellectual and conceptual influences of contemporary political economy in which the 'cultural turn' and analyses of the networks and legacies of social power have been particularly important in helping us to rethink overly productivist analyses of economic change. The fourth is the incredible deepening and revitalization that has occurred in Anglo-American economic geography over the past decade, particularly through its rich theorizations of localities, institutions, and practices. Across all four, our engagements have taken inspiration from the efforts of others to understand the cultural and political economy of post-socialism.[1]

Here, we address some of the practical, methodological and intellectual consequences of these engagements for the kinds of economic geography we pursue to understand the emergence of actually-existing post-socialisms. We stress the *plurality* of socialisms and post-socialisms to avoid any fixing of a normative history of complex social and economic arrangements, and the *global* nature of state socialisms and post-socialisms because the struggles over each have always meant a great deal to the form and practice of economies in other parts of the world. We pose several questions about the challenges post-socialisms create for the practice and methodologies of economic geography and radical political economy. Finally, we address some of the implications of our argument for methodological strategies, especially the kinds of collaborative research in which we engage.

Researchers and the geographies of encounter

Perhaps what first strikes you when you travel into CEE is the incredible mélange of practices, rhythms and identities that flow through particular places; past and present landscapes seem literally to tumble over each other suggesting that something new is underway, something old is being sustained, and something that combines the two is emerging. State socialist and market economies are articulating and re-articulating with one another in a heady mix of creative destruction and social transformation.

For many, the end of communism brought an opportunity to embrace a revitalized global neo-liberalism and the new economic and political opportunities it portended. This 'global gamble' (as Peter Gowan (1995) called it) was part of a political economy that exceptionalized the state socialist development model and re-wrote the past in terms of linear models of modernization, modernity and the natural ascendancy of capitalist liberal democracy (Smith 2002). In this sense, Fukuyama (1992) was only the most celebrated among many who saw in post-socialism an 'end of history' and a resolute and irreversible shift 'from plan to market' (World Bank 1996). In this view, the end of communism ushered in a period of great hopes of a return to Europe; a return that would expand economic opportunity and open personal and political freedoms (Pickles 2005).

Actually occurring transitions across the region achieved both goals, but only in part. Political liberalization generated intense struggles over economic resources. For those who were well positioned in relation to the networks of *nomenklatura* power, these new opportunities resulted in the amassing of vast wealth, increased mobility, the revitalization of urban landscapes of consumption, the flourishing of an urbane cosmopolitanism-cum-gangsterism, and a growing disregard for the conditions of the disadvantaged. For others, it meant job loss, neighbourhood infrastructural decay, increasing poverty, loss of social cohesion and safety, and a heightening of economic and personal insecurity. For those in the countryside, transformation generally meant rapid and unregulated de-development, the collapse of bus services, closing of day-care centres, uncertain healthcare provision, and mass unemployment (Meurs and Ranasinghe 2003).

More recently, with accession to the EU, border posts are disappearing, customs and immigration regimes are less threatening (at least to those with official papers), and public facilities for transport, accommodation, and other amenities are now well established. Mercedes and BMWs zoom past the horse-and-cart of the small farmer or Roma tinker. And television models new social identities for the emerging citizens of a larger Europe. This is a post-socialist landscape of montage; the materialization of super-imposed and re-worked development projects.

Sixteen years after the 'Velvet Revolution' many of its initial goals have been achieved: liberal, representative democracy, marketized economy, state withdrawal from civil society, and an opening to Western Europe. European integration has increased average incomes, labour migrations and remittance flows have expanded, and foreign direct investment and pan-European trade liberalization have created regional production networks in virtually all sectors (Pickles and Smith 2005). National and regional income levels have, however, yet to return to 1989 levels in many countries, wages in many of the new assembly industries are low, and unemployment remains chronically high. Transformation, integration and decline have produced new rounds of combined and uneven development, and it is these processes of socio-economic and regional differentiation that have exercised us throughout this period.

John's engagement with post-socialisms began as a result of research on late apartheid and post apartheid racial capitalism. In the late 1980s he participated in a series of exchanges between Hungarian and US, and Bulgarian and US scholars. At the time of the visit to Bulgaria, the Zhivkov government had made a fatal political miscalculation in turning once again to nationalist and ethnic politics to resolve deep economic problems. It expelled over 460,000 Muslim Bulgarians, mainly Turkish-speaking Bulgarians from the eastern part of the country (see Pickles 2001). The experience of witnessing these families moving their goods to Turkey (much like the displacements and removals in South Africa) has continued to shape every question he asks. The issue was immediately sharpened by the outbreak of hostilities in former Yugoslavia and the rapid escalation to full-scale wars and genocide. Why Bulgaria after 1989 did not come even close to such conflict became a central research question (Begg and Pickles 1998). In subsequent projects in CEE he has worked closely with colleagues to develop collaborative models of research focused on four issues: the geographical and economic consequences of state violence and nationalism under state socialism; the political economy of data and data institutions; the role played by neo-liberal models of 'transition' and 'reform'; and the ways in which state socialist legacies have been re-worked (through social networks, environmental policies, and economic internationalization) through what Karl Polanyi (1944) described as the counter-tendencies generated in all market economies towards social movements of reform and protection (e.g., Begg and Pickles 1998; Mikhova and Pickles 1994; Paskaleva et al. 1998; Pavlínek and Pickles 2000; Pickles et al. 2002, 2006).

Adrian's engagement with post-socialisms began while living and teaching in central Slovakia immediately after the collapse of state socialism. The opportunity to spend two years in a relatively small village close to the heart of the Soviet military-industrial complex provided for a fascinating lived experience. All around, the 'conversion' from military industries created mass unemployment and various re-workings of everyday life – not least the continued central role of household and community resources to provide basic household sustenance. Networks of integration and structures of assets that partly formed social and economic life during state socialism were becoming used afresh to provide lucrative opportunities for some – especially former managers – in their adjustments to the 'transition to capitalism'. At the same time, others were drawing on such legacies in their struggles to survive increasing unemployment, inequality and state retrenchment (Smith 1998, 2000, 2002). Since this time he has maintained a central interest in trying, both individually and collaboratively, to understand the grounded, local transformations of everyday life, the comparative experiences of regional transformations, and developing a critical, culturally-inflected political economy of post-socialism. Sitting, eating dinner, drinking and talking with friends back in the early 1990s, revealed the complexities of these lived experiences of marketization and liberalization, which continue to be central to his subsequent work.

Post-socialism and area studies

Both of us turned from development studies to research on state socialism and post-socialism precisely at the time when Cold War Soviet Studies (what is often called 'Sovietology') was breaking down (Pickles and Smith 1998). Partly as a result of this timing, our work has always been concerned to question these legacies of Cold War Sovietology. Like 'Orientalism' more broadly, these legacies have so deeply conditioned the history, content, and practices of research in area studies and we wanted no part of this. Our goals were and are to remain sceptical of 'Eastern' and 'Western' area studies that see regions as exceptionalist and essentialist.

For both of us this has meant being attentive to local actors and institutions. In seeking to respond to neo-liberal arguments our first goal has always been to take seriously the insights and needs of our colleagues in CEE. In the Soviet academy, researchers were trained across the network of national academies and universities, especially in Moscow. Individual scholars generally operated within a broad internationalist framework, often collaborating extensively on projects with colleagues from other Council for Mutual Economic Assistance (CMEΛ) institutions. The Institutes of the Academies of Science were able to gain access to state institutions and their data, and in some cases were the collectors of that information. Scholarship was rigorous, analytical, and data-driven, although – at the same time – it was also heavily dominated by the social hierarchy of the academy's own *nomenklatura* class.

Our second goal has, therefore, been to be attentive to the social struggles within the academy. For example, John's first encounter with these tendencies within state socialist science occurred at a conference in Morgantown, West Virginia, in which, after listening intently to US-based scholars and their own of cohort of colleagues, younger Bulgarian scholars immediately began chatting, walking around, or leaving the room as senior academicians presented their papers. Our initial response was one of intense irritation and frustration at this rude behaviour. Our astonishment grew when, after a long introduction in Bulgarian, the Bulgarian translator rendered the previous three or four minutes of presentation in English as 'He didn't say anything important' and, after the next three minutes of presentation, 'Again, he didn't say anything'. This was an editorializing that seemed perfectly normal to the younger Bulgarian scholars and continued intermittently throughout the conference. The struggles over *nomenklatura* power were being played out right there and it took the form of a fundamental shift in ideology and science: Marxism-Leninism was the required ideological superstructure for a *nomenklatura* class increasingly losing touch with the practices of their junior researchers. The senior scholars had fallen prey to what their colleagues judged to be an empty rhetoric of platitudes, 'statist' prescriptions, and uninspiring analysis.

Third, we have attempted to be attentive to our own myopias and commitments. In very concrete ways we were all schooled directly and/or indirectly in the traditions of Sovietology and Cold War social science with their own commitments to what Theodor Adorno called 'positivism in general' and a scepticism towards certain types of theory, particularly theories of totality, negation, and critique. As a result, area studies research was often a-theoretical, data-driven, and focused on the specificities of the region. In both 'East' and 'West' there was an inability or unwillingness to engage dialectically and historically with the big events then underway. In struggling to move beyond these traditions of Cold War and area studies/Sovietology it became crucial to think differently about the practice of research that we were undertaking. As one of us has written:

This [project] is first and foremost NOT a story of centred transition, or of economies outside the formal economy, or even of a 'retreat to' household economies, relations of reciprocity, informal exchanges, or what Caroline Humphrey and Ruth Mandel (2002: 1) have referred to as: 'the diversity of quotidian market activities of the people living in the new states and regions of the former Soviet Union and its eastern European satellites' – although it includes all of these. It is, instead, a narration that seeks to disseminate 'economic' subjectivities, not in the sense of multiplying them (although this certainly helps), but in a deconstructive sense that challenges us to locate economic identities and subjectivities in a diversity of post-socialist contexts, rhythms, and articulations. (Pickles 2004: 3–4)

Such a project is, in large part, about the mobility and mediation of ideas and presupposes an ethics of listening and conversation that has important methodological implications (Pickles 2005). In thinking through this ethics of listening and conversation (only briefly here), we begin by asking about the ways in which theory has or has not travelled.

Travelling theory

In one very important sense, 'theory' has travelled widely and to great effect through those concerned with market transitions and neo–liberalism. Post-socialism has been firmly centred on a neo–liberal moment and has been orchestrated by neo–liberal intellectuals and institutions.

In Slovakia, for example, a small group of neo–liberal thinkers associated with the Hayek Foundation became influential during the early years of this decade in debates over social assistance reform, tax policy and the more general process of liberalization (Smith 2007). Networked into the global neo–liberal think–tank system, such ideas had a transformative impact on a state open until elections in 2006 to embracing liberal market ideologies. More generally, post-socialist economic transformations elsewhere have occurred under the watchful eyes and guidance of country specialists at the World Bank, the International Monetary Fund, European Bank for Reconstruction and Development (EBRD), and associated agencies. In this sense, neo–liberalism in the region was constituted through the shock therapy policies of international institutions, resulting in what Guy Standing (2002: 51) has called the first revolution to be led by international finance capital and international financial institutions.

The power of global neo–liberalism appears, therefore, to have structured thinking about post-socialism in more powerful ways than have social democratic development models and critical political economy. Despite a few terse exchanges, critical Western scholarship appears to have had relatively limited direct impact and reception throughout the region. With only few exceptions, discussions of Marxism, state-led development projects, and theories of collective economic action hold little interest for most scholars in the region.

Here, however, we want to focus on some alternative readings of this narrative of transition and suggest that more critical theory rather than less is emerging. In one very simple way, we want to deconstruct the notion that there are fundamental divides between progressive scholars in the West (social democratic and fundamentally opposed to the neo-liberal project) and those in the East (democratic and fundamentally opposed to any wholesale and un-dialectical rejection of neo-liberalism). Both sides are misconceived and can be thought quite differently. How we read post-socialism thus entails the parallel and related question of how we read neo-liberalism. After the Cold War we are all, in effect, post-socialists.

Some voices from the region have suggested precisely this (Žižek 2001). For Žižek the problem lies in attempts to write post-socialism solely as a story of CEE. The geopolitical deployment of Cold War discourses of 'totalitarianism' by the 'West' directed thinking about communism inward into the region. As a result, the crucial importance of discourses of totalitarianism in 'Western' as well as 'Eastern' geopolitics and political economics was overlooked. Post-Cold War thinking about post-socialism cannot re-inscribe this exceptionalism. A renewed historical imagination is required, particularly one that interrogates the parallel totalitarian impulses in the logics and discourses of transition and market-idolatry, and sees in soviet-style socialism much more complex and nuanced social relations and political conditions.

While the discourses of totalitarianism may appear to be an overly abstract entry point into thinking about post-socialist research methods and approaches, the issues raised by Žižek are, in fact, very practical concerns. How does the 'visitor' carrying out research in the region understand and represent state socialism? This is important because it can have immediate consequences for the kinds of social relations and research access that are possible. Precisely how neo-liberalism is being interrogated by the 'visitor' is immediately apparent to scholars, government officials, trade unionists, managers, and translators alike. Basic credibilities are won and lost very quickly in this game of translations.

For many years we have participated in meetings and read papers written by technocrats, scholars, and conference-goers deeply committed to forms of social solidarity and positive post-socialist transformations, only to hear (from the left and the right) the same flattening of socialist history (as 'lost hope' or 'terrifying gulag') and the rendering of state-socialism through the very categories of neo-liberalism (ill-conceived, inefficient, even corrupt). Such readings have consequences for the ways in which policy is made, but more importantly they also reject the complex dialectics and articulations that frame the present conjuncture and, as a result, they shape and limit the kinds of collaborative relationships that are possible.

Mediating the legacies of Stalinism and state socialism is a difficult task, not least for researchers wishing to build a political economy of post-socialism informed at least partially by Western Marxism. Strong Marxian critiques have emerged from the region, most notably from Boris Kagarlitsky and Slavoj Žižek, and Michael Burawoy has carefully and systematically argued for the greater analytical power of classical versus neo-classical sociologies. But these efforts remain minoritarian, with a larger readership outside the region than in it and, in the case of Kagarlitsky and Žižek at least, with only a handful of readers within economic geography. Analyses of class formation, property forms, and the role of the state continue to figure prominently in the training and research of established academics, politicians, and trade unionists in the region (e.g., Crowley and Ost 2001). Some younger scholars schooled in 'Western' traditions of critical human geography are engaging with ideas and approaches not conventionally seen to be part of geography in post-socialist regions. Issues of gender, sexuality, and feminism, for example, are particularly important in this regard.

Given these various ways in which ideas travel, it might be useful to ask about the ways in which researchers in the region also convey a not so hidden set of ideas about post-socialism. Certainly everyone seems to have opinions about the causes of and remedies for the problems of transformational economies. From corruption to inefficiency, to lack of know-how, to a lack of a history of democracy, to personal or cultural traits that favour demagogues, Western social scientists have analyzed and characterized central and east Europeans in a wide variety of ways, and many of them are easily recognized and sometimes shunned by residents throughout the region. Chakrabarty's (2000) analysis of the role of modernist and colonial discourses that consign other peoples and regions to 'the waiting room of history' may not be common currency among scholars in the region, but they well understand the hidden cultural politics of the Westerner 'telling' them about the real losses they have suffered, about the benefits of central planning, about the need to overcome the inefficiencies of over-employment, the need to make governmental institutions more transparent, and the need for workers to spend more time working more effectively and efficiently.

This unintended geopolitics of visitors armed with 'explanations' has inflected (and infected) the social relations of research throughout the region, and it has been compounded with the coming of the Eurocrats and their consultants and programmes. A whole range of administrative, legal and institutional reforms, translating EU norms into practice, have been implemented. The result has been a transformation of regulatory regimes 'from the outside' with agendas being set by EU institutions and, as with the shock therapists before, local scholars and policy-makers have become wary translators of reform (Smith and Hardy 2004). Here we provide two short examples.

First, and perhaps the sharpest in terms of the practical consequences of research, is the misreading of state-socialist industry in terms of inefficient factories, under-employed labour forces, and unskilled party managers. On one visit to a large Bulgarian petrochemical *combinat* we were shown into the control room. In it were nine workers, all men. Two were playing cards, one was asleep on a cot in the corner, three were drinking coffee and reading the newspaper, two were fixing a broken dial, and one was sitting at the control desk. Not surprisingly, the group of visitors was surprised by the number of workers in the room and by their apparent disinterest in the actual workings of the plant. There is, of course, much to be said about the problems with state socialist industrialization. Like this petrochemical factory, the socialist firm is all too easily interpreted as inefficient and poorly managed. But, such interpretations fail to understand the state socialist factory as a site of complex economic relations and exchanges negotiated over time by managers who lived and died by their ability to effect complex material and labour trades at key points in the five-year and annual planning cycle. These practices were necessary elements of a production politics predicated on shortage (Pickles 1995). Decisions about resource allocation had to include investments in 'non-productive' activities for maintaining the 'social wage' (such as vacation resorts, sports facilities, housing, and transport), creating – and in some cases still sustaining – intense local geographies of dependency. Of course, the control room of the petrochemical plant may have been over-staffed, under-attended, and operated more as a social service to the workers and the local community than as an efficient productive enterprise. But there was nothing in the encounter that allowed us to know this and there is a good deal of evidence to suggest that something else entirely was going on in this and other enterprises across the region.

A second example illustrates further the importance of rethinking totalitarianism and 'blocky' concepts regarding state socialism. While the geopolitics of the 1980s saw a hardening of US attitudes towards the USSR and military escalation around its borders, as early as 1983 several factories across CEE, most notably in southeastern Bulgaria and east Slovakia, were producing military uniforms in state-run factories working under state planning regimes contracting for NATO, illustrating the need to think more seriously about the relational geographies of state socialism and post-socialism. Throughout the 1980s emergent forms of economic integration between 'East' and 'West' in Europe set in place social and economic relations that have since been mobilized to consolidate, for example, apparel export production (see Pickles et al. 2006). These networks and linkages have been transformed as asset structures on which to build new forms of production relations and new forms of geopolitical integration in a post-enlargement EU.

Embedded research and the cultural politics of collaborative research

In both of these cases, the researcher begins with a set of assumptions about state-socialist and post-socialist economic practices only to find that quite different explanations are necessary. But how are such 'different explanations' to be derived? This is, of course, a difficult question and involves many important issues of research methodology. But it also involves some basic issues of politics and attunement in research practice. We conclude with a brief discussion of the implications of these issues for collaborative research.

Central to both our individual and collective projects has been a long-standing concern for the politics of our research collaborations. These have required that we negotiate very different conceptions of research practice. As a result, understanding actually-existing post-socialisms requires thinking about and managing the enormous social and conceptual commitments that come with collaborative research. In part, this need is very practical. Take, for example, the issue of data and information.

Every researcher working in the region soon begins to understand the 'political economy of data institutions'. There are real differences in the way in which data collection and circulation have shaped and continue to shape the concrete practices of research in and about the region. Under socialism data infrastructures were well established, and reporting through the channels of the party state and its administrative structures was required of all functionaries. The resultant information and data were carefully guarded under penalty of law and access was highly controlled. In the real economy where material goods and labour were in short supply and where hoarding and trading were essential practices of everyday life, informal 'data' trading was common. Where information meant social power or tradable resource, it too circulated through social networks or at a cost, in turn shaping the kinds of issues it was possible to study. Access thus conferred social power on party members, bureaucrats, and scholars, a situation compounded by the loss of state funds after 1989 and a growing dependency on foreign-funded projects became relatively more important. In some periods and places, such foreign-funded projects were the only sources of funds for fieldwork and travel, as well as for income supplements for local researchers or for graduate student support.

These practices of information provision and institutional funding shaped research practices and they continue to be important in the economic uncertainties of post-socialism. Managers continue to complete firm surveys because the request comes from influential state institutions, such as the Academy of Sciences, or because personal networks have been mobilized, and research institutions continue to depend on contract research with international organizations and scholars. In recent years new laws of reporting, transparency requirements, and the growing importance of codes of conduct and international standards in business contracting have led many managers, government officials, and others to adopt more open data- and information-sharing practices and these have been encouraged by changes in funding behaviour. Research funds remain limited and this has forced research institutions to continue to depend on foreign projects. However, in a funding landscape increasingly dominated by EU sources, projects now explicitly require the building of scholarly networks and the implementation of data-sharing practices.

Elsewhere we have described the ways that social relations (*bliski* networks) function in state socialist and post-socialist economies as resources to sustain livelihoods in conditions of shortage (Begg and Pickles 1998; Smith and Stenning 2006). While we study these networked social relations of trust, reciprocity, and access in the form of enterprise restructuring strategies and in the construction of household livelihood strategies, we also live these networks through our own research practices. Fieldwork requires that we first visit with friends, organize social gatherings, and return courtesy calls with bureaucrats, managers, trade unionists, and colleagues. These are expected. It is also a responsibility and one of the central pleasures of ongoing scholarly collaboration, and it creates dense structures of reciprocal obligation. These obligations are not only important rituals through which one is able to gain institutional access or access to interpretation and information, data, and contacts, they are also responsibilities and pleasures that come with friendship and they are times of vital scholarly translation.

NOTE

1 Of particular importance have been, among others, Michael Burawoy's politics of production and comparative transitions, Ivan Szelenyi's rethinking of class power, Caroline Humphrey's elaboration of post-socialist everyday economies, Katherine Verdery's ethnographies of symbolic and identity politics, Chris Hahn's focus on common property resources, Gernot Grabher and David Stark's path-dependent nature of post-socialist transformations, Gerald Creed's ethnography of the everyday appropriations and domestications of socialist practices, and Mieke Meurs' institutional and regulatory analysis of agrarian transformation.

Section 3

Quantity and Quality: Beyond Dualist Economic Geographies

13 HYBRID GIS AND CULTURAL ECONOMIC GEOGRAPHY

Mei-Po Kwan

Predisposition

In the late 1980s I conducted a study on the relationship between economic development and government technology policy in the (then) newly industrializing countries of Asia. The study adopted a qualitative case study approach, which involved archival research and fieldwork that included interviews with government officials responsible for development and technology policy. The project was for my Master's thesis with the Graduate School of Urban Planning at UCLA. It was informed by the work of my thesis advisers (Michael Storper and John Friedmann) and several geographers at UCLA (e.g., Susan Christopherson, Ed Soja and Allen Scott). After I filed my thesis, my path underwent an intriguing turn. I went to the University of California, Santa Barbara, for my doctoral studies with a focus on GIS (Geographical Information Systems).

When my academic career began in 1995 at the Ohio State University, my research focused mainly on the development of GIS and quantitative methods for understanding the activity travel behaviour of individuals. Like many others in their early careers and on the way to tenure (as articulated by David Rigby in this volume), I faced the pressure to publish in top geography journals and to obtain external funding for my research, both were highly challenging in the intellectual environment of the discipline in the mid-1990s. By that time the cultural turn in geography in general and in economic geography in particular had led to a decisive move towards post-positivist, anti-essentialist, reflexive, context-sensitive and qualitative research practices (Barnes 2001; Crang 1997; Peck 2005; Thrift 2000). New theoretical and methodological perspectives (e.g., feminist, post-structuralist and postcolonialist) had become much more influential, while new metaphors and new subject matter had been invoked in attempts to create new strands of geographic inquiry, such as the new economic geography (e.g., Barnes 2001; Ettlinger 2001; Gibson-Graham 1996; Hanson and Pratt 1995).

This was a time when the credibility of GIS and quantitative methods was being seriously questioned (e.g., Taylor 1990). In an attempt to initiate an independent post-doctoral research programme that might address some of these concerns, I became interested in feminist geographers' research on gender, work and commuting (e.g., England 1993; Hanson and Pratt 1988). Drawing upon my training in economic geography, critical social theory and qualitative methods, I mapped out a research agenda that amalgamated elements of feminist geography, economic geography, urban and transportation research, GIS, and quantitative methods.

During this period, my research programme has undergone considerable transformation in both its substantive focus and methodological orientation. This transformation stemmed partly from the need to negotiate the tension brought about by the cultural turn in geography and the critical discourse on GIS, and partly from my recognition of the limitations in conventional GIS and quantitative methods for grappling with the complexity and richness of lived experience. By the end of the first decade of my academic career, I had become a geographer whose primary research interest is in the gendered, raced and classed geographies of everyday life, and whose GIS practice is shaped more by post-structuralist-feminist perspectives than by analytical geography as conventionally understood – although my work still speaks to the interests of many quantitative geographers.

The narrative that follows portrays my often tortuous negotiations of the theoretical and methodological tensions brought about by the cultural and qualitative turns in geography, as well as the critical discourse on GIS. I reflect upon my move towards more qualitative and interpretative modes of analyzing geographic data, and my decisions underlying this shift. I explore issues pertinent to the use of GIS in economic geographic research. Through this reflexive narrative, I show how my GIS practice and the intellectual milieu within which I have worked have been intertwined and mutually constitutive.

Gender, occupational status and the journey to work

The research project that I focus on in this chapter concerns the complex relationships among women's employment status, journey to work and space-time constraints (Kwan 1999). I formulated my research questions in light of two important trends in women's participation in the labour force and the gender division of social and domestic labour. First, as more women participate in the labour force, some have been able to achieve relatively high occupational status and income. Second, as the proportion of women who can use their own automobile to commute increases, many women now have better spatial mobility to work farther from home than before. Many researchers believe that these two trends together will enable certain groups of women to take up high-status jobs farther from home, and this in turn will lead to a decrease in their shares of domestic responsibilities within a household. Smaller shares of women's household responsibilities, according to this view, imply that men will take up larger proportions of household responsibilities. These changes in the position of women in the local labour market and in the gender division of domestic labour in turn will likely lead to more equitable gender relations within the household and society at large (Gershuny and Robinson 1988).

I examined the claims of this so-called convergence hypothesis, which argues that women's improved occupational status would 'naturally' lead to progressive change in the gendered division of social and domestic labour. I also examined whether the constraints associated with women's disproportionate responsibility for domestic tasks are still important in determining their employment status and commuting distance. The main theoretical constructs I used were a set of time-geographic concepts employed in earlier research on the geographies of women's everyday lives (e.g., Tivers 1985). 'Space-time constraints' stem from two main sources: the first is the limited time available for a person to perform various activities in a particular day (the time budget constraint); the second arises from the fact that activities that need to be performed at fixed location or time (e.g., childcare drop-off) restrict what a person can do for the rest of the day (the fixity constraint). Past studies observed that space-time constraints significantly affect women's job location, occupational status, and activity patterns (e.g., Hanson and Pratt 1995).

I chose a quantitative approach because the literature on women's labour market position and commuting is largely quantitative, and the primary intention of my study was to overcome some of its limitations. For instance, no previous studies had attempted to quantitatively measure the space-time constraint associated with the performance of various daily activities and evaluate the extent to which they affect women's employment status and commuting distance. In addition, the focus was often exclusively on the distance (or travel time) between home and workplace and its relationships with women's employment status, with scant attention to the location of non-employment activities relative to a person's home and work sites. As many of these non-employment activities are associated with the performance of specific household responsibilities, shaping job and residential choice in significant ways, I felt that an explicit focus on these activities might reveal important gender differences within a household.

Quantitative analysis of quantitative data

To address the complex relationships among women's domestic responsibilities, employment status, journey to work and space–time constraints, I decided to collect quantitative data from a sample of working women in Columbus (Ohio, USA), using activity-travel diaries and to utilize statistical techniques (e.g., structural equations model) to analyze the data.

I had to resolve several operational issues about how to turn the notion of space–time constraints into something measurable before collecting the data. I solicited information about the space–time fixity of each activity a person performed through an activity-travel diary survey. The diary recorded details of all activities and trips made by the respondent in two designated travel days. The survey instrument comprises two parts: a household questionnaire and a two-day activity-travel diary. Using this survey instrument, I collected an activity-travel diary data set from a sample of adults (over 18 years of age) in households with one or more employed members in the study area in 1995. The household questionnaire collected information about the socio-economic characteristics and transport resources of all household members. The diary recorded information on activities and trips (including street address, travel mode, car availability, routes taken, the primary purpose of each activity, a subjective fixity rating for each activity, and other individuals present when performing each activity).

Because the small number of ethnic minorities in the sample does not allow for meaningful statistical analysis, they were excluded from the analysis (this would not have been the case if qualitative information had been collected). The final sub-sample consists of three groups of white people: 28 full-time employed females, 13 part-time employed females (who work less than 35 hours a week), and 31 full-time employed males. I analyzed the differences in fixity constraint experienced by individuals of these three groups using simple descriptive statistics and analysis of variance (Kwan 2000). The results show that women employed part-time encounter more fixed activities in their daily lives than the other two groups. Many of these fixed activities are associated with household needs that have a strong restrictive effect on the locations of their out-of-home activities and job location (such as picking up a child from a day-care centre). Further, despite the fact that women employed full-time travel longer to work than men, they experience a higher level of fixity constraint than men. This result is surprising considering the high occupational status and high level of access to private cars of the full-time employment women in the sub-sample. It contradicts the convergence hypothesis, which suggests that women's improved occupational status would lead to more equitable gender division of domestic labour and lower level of fixity constraint.

Qualitative analysis of quantitative data

In light of these unexpected results, further in-depth causal analysis of the activity diary data using statistical methods seemed necessary for unravelling the complex interactions among gender, employment status, journey to work and space-time constraints. But analyzing these interactions is difficult because human activities and movements in space-time are complex trajectories with many interacting dimensions. These include the location, timing, duration, sequencing and types of activity and/or trip. This characteristic of individual activity-travel patterns makes the simultaneous analysis of its many dimensions and their interactions with gender or employment status difficult.

In addition, the trajectories representing human activities and movements in space-time unfold in the concrete geographical context of particular places, while few if any conventional quantitative methods satisfactorily take such contexts into account.

First, few of these methods can handle real geographical locations of human activities and trips in the context of a study area. Often, the spatial dimension is represented by some abstract measures derived from real geographical locations (e.g., distance or direction from a reference point, such as home or workplace). Further, locational information of activities and trips was often aggregated with respect to a zonal division of the study area (e.g., traffic analysis zones or census tracts). With such zone-based data, journey to work or commuting distances estimated using zone centroids may lead to considerable errors and a considerable amount of important information in the original data is lost. Second, the small number of subjects that belong to certain social groups (e.g., ethnic minorities) in a sample often preclude the possibility of meaningful statistical analysis. They are often treated as outliers or excluded from the analysis.

Third, given the complexity of activity diary data, having a preliminary understanding of the behavioural characteristics or uniqueness of the individuals in the sample at hand can often lead to more focused and fruitful methods or models in later stages of a study. Effective methods for exploring these data are necessary for avoiding errors in model specification or taking into account any behavioural anomalies in the data set. Fourth, since many statistical methods used in previous studies are designed to deal with categorical data, organizing the original data in terms of discrete units of space and time has been a necessary step in most studies in the past (e.g., as in the case of log-linear models). Discretization of temporal variables, such as the start time or duration of activities, involves dividing the relevant span of time into several units and assigning each activity or trip into the appropriate class (e.g., dividing a day into eight or twelve temporal divisions into which activities or trips are grouped). Discretization of spatial variables, such as distance from home, involves dividing the relevant distance range into several 'rings'. Since both the spatial and temporal dimensions are continuous, results of any analysis that are based upon these discretized variables may be affected by the particular schema of spatial and/or temporal divisions used. The problem may be serious when dealing with the interaction between spatial and temporal variables since two discretized variables are involved.

Conventional quantitative methods have many limitations, even when applied to analyze quantitative data. Because of the data requirements of particular analytical procedures, much of the variation in the original data set may be seriously distorted before the analytical phase. As it was problematic to proceed with standard procedures, I started to consider other ways for dealing with the data in order to avoid some of the problems discussed above. This led me to consider GIS-based 3D geovisualization as a means to conduct preliminary analysis of the data. Visualization is the process of creating and viewing graphical images of data with the aim of increasing human understanding (Hearnshaw and Unwin 1994). It is particularly suitable for dealing with large and complex data sets because conventional inferential statistics and pattern recognition algorithms may fail when a large number of attributes are involved (Gahegan 2000). For instance, spatio-temporal data can be explored before they are discretized for further analysis or modelling using visualization. Geovisualization (or geographical visualization), on the other hand, is the use of concrete visual representations of the relevant geographical elements to make spatial contexts and problems visible (MacEachren et al. 1999). This facilitates the identification and interpretation of spatial patterns and relationships in complex data in the geographical context of a particular study area.

Since GIS-based geovisualization takes GIS images as its primary input and does not involve quantitative analytical procedures, I recognized the role of GIS as a useful tool for the qualitative analysis of quantitative data. I developed several GIS-based 3D geovisualization methods for visualizing the activity-travel diary data I collected for the study. These include space-time paths, space-time activity density surfaces, and space-time aquariums. With the help of these GIS-based methods, the interaction between the temporal dimension and the spatial dimension in structuring the daily space-time trajectories of individuals was clearly revealed. I was able to detect differences among the three groups of subjects with respect to their daily activity patterns and commuting distances.

For instance, the space-time paths of the 28 women employed full-time reveal their long work hours and their tight time-budget constraints, which allow for only a few non-employment activities to be undertaken throughout the day. The space-time paths of the women employed part-time, on the other hand, are more fragmented, as they tend to perform more non-employment activities during the day given their less restrictive time-budget constraint. Further, the pattern of the standardized space-time paths reveals that non-employment activities for women and men working full-time are strongly oriented along and constrained by the home–work axis. Because of the time-budget constraint imposed by their work hours, a considerable proportion of their non-employment activities are undertaken in the evening. During the day, not only are there few non-employment activities, but the distance of these activities from the home–work axis is also highly constrained.

These results obtained from the qualitative geovisualization of the data enabled me to formulate more informed and realistic quantitative models in later stages of the study. For instance, I conducted a canonical correlation analysis and found that women encounter more fixed activities than men in the day-time regardless of their employment status, and women who work part-time experience higher levels of day-time fixity constraint than the full-time employed women and men in the sub-sample. Further, the level of day-time fixity constraint one experiences depends more on one's gender and whether there is someone in the household who can and is willing to take up some of the household-serving activities in the day-time. If there is someone to share the household- and child-serving activities within the household, the burden of fixity constraint a woman faces in her daily life will be greatly reduced. The most significant implication of this is that redressing the domestic division of labour and gender relations within the household may improve women's labour market position and income-earning potential.

To analyze the complex interrelations among women's day-time fixity constraint, non-employment activities, household responsibilities and employment status, I estimated a non-recursive structural equation model with latent variables for the women in the sub-sample (Kwan 1999). I found that household responsibilities, besides exerting a direct effect on women's employment status, has an indirect impact on it through the mediating effect of day-time fixity constraint and the number of out-of-home non-employment activities. Increases in women's day-time non-employment activities due to part-time employment tend to heighten the level of fixity constraint, as part-time employed women have the time and flexibility to undertake more fixed activities during the day. These results confirm that fixity constraint has a significant impact on women's employment status (where women with higher levels of fixity constraint are more likely to work part-time).

These findings call into question the belief that increasing female participation in the labour force will lead to significant change in women's gender roles and space-time constraints. They also suggest that the situation of women may not change much without first challenging gendered divisions of domestic labour. Despite the belief that recent trends in the increasing number of women with higher occupational status and improvement in their access to private means of transportation will lead to changes in traditional gender roles, the results of my study suggest otherwise.

GIS as qualitative method in economic geography

The GIS methods I developed in the study were used to visualize the space-time paths and activity patterns of the survey respondents without first reducing the original data to statistical aggregates. They not only helped to retain the particularities of each individual subject, they also rendered the interpretation of the results less disembodied, as the meaning of each of the paths has to be understood in the specific socio-spatial context of the life of each subject. I also brought to the interpretative process my personal knowledge of the study area, my feminist sensibility and my distinctive GIS methods. As I also talked to some of the subjects over the phone to clarify issues and obtained valuable insight about the processes through which domestic division of labour is negotiated, the subjects were given some (albeit limited) 'voice', instead of being represented entirely by the quantitative data they provided. I recognized that qualitative information is also crucial for shedding light on these processes (which are not apparent from the quantitative data). For instance, it was impossible to infer the power relations between men and women or the consequences for the distribution of household responsibilities among household members. It would be more fruitful for this kind of study to employ a mixed method strategy that combines both quantitative and qualitative methods, such as the one used by Hanson and Pratt (1995).

I decided to pursue more qualitative ways of using GIS in geographic research in my subsequent studies, in order to convey more of the complex lived experience of research subjects within GIS environments. Although GIS can only handle digital information, I recognized that the recent development of digital technologies has greatly expanded the kind of information GIS can deal with. Contemporary forms of digitization now include a wider array of representational possibilities than merely numerical or quantitative data. For instance, qualitative data such as digital photos, video and voice clips can be linked or incorporated into GIS. In studies using qualitative methods, subjects' handwriting, hand-drawn maps and other sketches collected through in-depth interviews and other ethnographic methods can also be incorporated in a GIS. The use of GIS therefore does not necessarily preclude the use of rich and contextualized primary data of subjects or locales, or the possibility to incorporate multiple views of the world or forms of local knowledge (Kwan and Knigge 2006).

For instance, I used mixed methods in a recent study of the impact of anti-Muslim hate crimes on the geographies of everyday life of the Muslim women in Columbus (Ohio, USA) after the terrorist attacks of 9/11. Fear of being harassed or attacked has become part of the daily lives of American Muslims as they recognize the threat suggested by numerous reports of violent anti-Muslim hate crimes in the media and on the internet. Muslim women are especially vulnerable to anti-Muslim hate crimes and discrimination, as they can be easily identified in public places, due to their distinctive religious attire – most women of the Islamic faith wear the Muslim headscarf (*hijab*) in public spaces and in the presence of men outside the family. Muslim women are vulnerable also because their traditional gender role in the family renders it necessary for them to undertake many out-of-home activities in their normal daily lives. As many of these household responsibilities impose restrictive space-time constraint on their daily lives, the need to undertake them can make their lives particularly perilous and stressful after 9/11.

In the study a specially designed activity-travel diary was first used to collect activity data of the subjects on a survey day after 9/11. Part of the diary also requests respondents to provide data on their activities on a typical weekday before 9/11. The qualitative part of the data collection phase consists of a combination of map sketching and in-depth interviews, both are conducted at the subject's home. Research participants not only responded to the semi-structured questions during the interview, they also drew on a map of the study area as they were telling the story about their personal experience shortly after 9/11. The interviews were tape-recorded and transcribed into text for later analysis. In addition, photos of several selected subjects were also taken with their permission and used in the qualitative GIS analysis in a later phase of the study.

I developed several qualitative GIS methods to represent these data within a 3D GIS environment. One is the daily space-time path of each of the subjects, which is the trajectory that traces their movements over space and time as they undertake their activities and travel in the study area. It portrays the subject's personal experience in space-time. These paths are colour-coded to reflect the level of fear and perceived danger experienced by the subjects as they recalled during the interview. Text fragments from the interview transcripts, audio clips and photos were linked to relevant segments of these 3D space-time paths in the GIS. With a considerable amount of contextual geographic data incorporated into the 3D GIS environment (such as all the buildings and the transport network), an interactive environment was created in which I visualized and explored the personal experience of the subjects through a reconstructed 'visual narrative'. Another method I developed may be called '3D GIS videography', involving the creation of movies by capturing and assembling still 3D GIS images that represent the person-specific experience and story of a particular research participant (Kwan 2007). For instance, a movie on the 'landscape of fear' that I produced represents a Muslim woman's experience as she travelled along a major road in the study area. In the moving scenes, the buildings along the road were colour-coded to indicate the level of perceived danger experienced by the subject. Using these 3D GIS methods, I was able to visually represent the stories and limited activity space of many Muslim women after 9/11.

These examples not only suggest that GIS can be a useful method for illuminating certain aspects of women's everyday lives. They also indicate that GIS can accommodate modes of analysis other than conventional quantitative or spatial analytical methods (such as the interpretation of visual representations of quantitative or qualitative data). Since the early days of its development, GIS has been understood as a tool for the storage and analysis of quantitative geographical information. This understanding of GIS has underpinned much of the critical discourse on GIS in the 1990s, in which both GIS critics and researchers considered GIS mainly as an apparatus for positivist/empiricist science or quantitative methods. This debate has led to an understanding of geographical methods that places GIS at one pole of a series of binarisms – positivist/quantitative/GIS methods versus critical/qualitative methods, and GIS/spatial analysis versus social/ critical geographies (Kwan 2004). Understanding GIS merely as a quantitative or empiricist method, and placing it at the polar opposite to critical geographies or qualitative methods, forecloses many opportunities for economic geographers to use GIS in qualitative or mixed method research (Kwan 2002; Sheppard 2001). The representational potential of GIS can be used for enacting creative discursive tactics that disrupt the dualist understanding of geographical methods – where visual images (albeit generated and composed with digital technology), words and numbers are used together to compose contextualized cartographic narratives in geographical discourse.

GIS as hybrid practice in economic geography

As I began my career in the mid-1990s, little did I know that I would be undertaking mixed-method research in feminist economic and urban geography a decade later. The cultural and qualitative turns in geography as well as the critical discourse on GIS have shaped my research practice in important ways. My GIS method is now characterized by an esoteric hybridity, amalgamating elements that are conventionally understood as incompatible. I pay much more attention to the effects of my positionality and the politics of research, especially the power relations between myself as a researcher and those I study. My negotiation of the cultural turn in geography in general and in economic geography in particular made me realize that it is problematic to consider GIS methods as either qualitative or quantitative. Unlike quantitative methods, GIS can be adapted to handle different types of data and its nature is materialized only in the concrete contexts of particular applications. Even when the original data used are largely quantitative in nature, transforming these data into various types of visual representation (e.g., 3D scenes) allows, to a certain extent, a more interpretative mode of analysis than what conventional quantitative methods would permit. It is difficult to make a clear-cut distinction between quantitative and qualitative methods when using GIS. Perpetuating the binary constructions that emerged in the critical discourse in the 1990s will likely have a considerable negative impact on economic-geographic research in the future. In addition, certain GIS methods, such as 3D interactive geovisualization, offer a means to overcome some limitations of conventional quantitative and statistical methods. By retaining the original data without first reducing them to statistical aggregates or averages, these methods retain the particularities of individual subjects, while providing a general picture of their behavioural patterns at the same time. This will help destabilize the traditional distinction between the idiographic and the nomothetic approaches to geographic research.

14 EVOLUTION IN ECONOMIC GEOGRAPHY?

David L. Rigby

This chapter offers a brief sketch of the research programme that I have followed. It begins with graduate training in Marxian political economy and uneven development and extends to current interests in evolutionary models of technological change and regional economic dynamics. I spell out the choices made at different stages in my academic career, discussing the factors that influenced those decisions. These choices focused on research questions, on competing theoretical frameworks, and on evaluation of alternative methodologies. The choices were influenced by more practical considerations too. Initially, the concern was with choosing a research topic that was personally interesting and looked like it might have a shelf-life longer than my graduate career: I wasn't terribly successful! Later, the concern shifted to what I perceived to be the direction in which the academic 'market' for economic geographers was moving. Over the last decade or so that market has de-privileged what have become known as essentialist or meta-theories such as Marxist political economy, and it has railed, to my mind naïvely, against mathematical modelling and quantitative analysis.

I begin by outlining the theoretical and methodological divides that I encountered upon entering graduate school. These reflected the disciplinary concerns of the time and they played out in different ways, from stark choices between neoclassical economics and political economy, to quite heated discussions around the structure–agency debate. Those times were exciting. They also were a little unsettling, having immediate impacts on research funding and the composition of dissertation committees: in my case this was made a little more complicated when my adviser moved about as far away from me as he possibly could, just as I began the PhD. Following the moves of Michael Dear, Ruth Fincher and Michael Webber, the Department of Geography at McMaster University embarked upon its own phase of restructuring, one much less sympathetic to political economy.

Next, I discuss research strategy made as a junior faculty member with the usual concerns about tenure. That strategy was significantly influenced by the steady erosion of interest in Marxist political economy, and an increasing dismissal of the relevance of analytical modelling and quantitative investigation. Trained precisely in those areas, tenure looked very distant. These changes were difficult to comprehend. We clearly hadn't answered the questions that had been posed, yet many were ready to abandon those questions as fashions changed. Many political economic geographers, myself included, took refuge in 'mid-level theory' and related empirical analysis, producing work that was essentially agnostic in terms of connecting to more abstract visions of the capitalist economy. This agnosticism meant safety. Such work could, and generally did, ignore potential inconsistencies in more abstract models, it was not as susceptible to criticisms of reductionism or essentialism, and, frankly, it was easier to publish. On reflection, however, it seems that economic geography abandoned many of the big questions.

Finally, I discuss how my research programme shifted towards exploration of alternative, non-equilibrium-based models of economic change. This work has largely focused on evolutionary economics, and on attempts to understand aggregate, industry- or region-wide processes of technological change through the dynamics of the individual economic agents (plants) that comprise industrial or regional economies. This shift reflects changing interests, but also it is an acknowledgement of the difficulty of building a career on questions that have become more marginal within the subdiscipline, and of employing methods that fewer and fewer reviewers are to understand.

In search of a dissertation

It is often said that there are two kinds of prospective graduate students, those who know precisely what their research question is and those who have no idea. I certainly fell into the latter category. It would also be fair to say that on entering graduate school, though I had some naïve understanding of different philosophical traditions and competing theoretical claims, I was unaware of the academic wars raging over the future of economic geography, and of the disquiet that Marxian political economy had introduced to the social sciences more broadly. That all probably changed in my first week in the Department of Geography at McMaster University. This was 1982, and within economic geography, close to the height of the battle between adherents of 'traditional' modes of inquiry largely built around choice-based utility models and methodological individualism and adherents of political-economy-based approaches. This was also a period of significant economic turmoil. The long post-war economic boom had decidedly ended, with the early 1980s heralding the deepest recession most industrialized economies had endured in fifty years.

It is unclear precisely how the academic battles and the real economic crises of the day were related, though the rise of political-economic approaches throughout the social sciences surely reflected the material conditions of those years. Equilibrium models of unlimited economic growth were increasingly challenged by models of economic crisis, by long-wave and related theories of economic expansion and recession, by widespread empirical studies of industrial decline and job loss, by competing models of work organization, technology and institutional practice, and by the increasing reach of the global economy (Bluestone and Harrison 1982; Froebel et al. 1980; Massey and Meegan 1982; van Duijn 1983). My impression today is that economic geography is nowhere near as relevant as it once was, at least in terms of its impact on policy.

Geography at McMaster was an exciting place during this period, not least because of the quality of the faculty, but also because that faculty spanned widely divergent philosophical, theoretical and methodological positions. The composition of the graduate student body reflected those differences, but as clear harbingers of the future, those students were increasingly moving left into political-economy. This process accelerated as a group of graduate student 'refugees' in geography abandoned the University of Toronto, largely for political reasons, and enrolled at McMaster.

The human geographers were divided along at least two major fronts. One was the debate over agency, pitting behavioural geographers and economic geographers trained in the neoclassical tradition (and advocates of methodological individualism) against those who embraced the arguments of political-economy. Discussion was often presented in stark terms, as a choice between individuals maximizing utility and structural Marxism of an Althusserian variant that gave no voice to the individual (Chouinard and Fincher 1983; Duncan and Ley 1982). There were also real tensions between advocates of traditional, neoclassical economic theory, between those working on models of urban form and function, and others seeking to develop Marxian models of the evolution of the capitalist space-economy. Though the form of analysis (mathematical modelling) was similar for both camps, those models were based on quite different assumptions (Barnes 1987; see also Myers and Papageorgiou 1991).

On reflection, I was fortunate to have had the opportunity to pursue very different research paths with a faculty who were all both committed to their students and had a significant impact on the sub-field in which they worked. While those possibilities should have made the choice of dissertation topic difficult, in the end it was relatively easy. There was an initial pull to work on the impacts of noise pollution around highways. While certainly concrete, and perhaps more importantly, funded, this project really didn't seem to be that exciting, especially given the debates in which most of my colleagues were immersed. That raised questions about the possibility of publishing, of selling this work on the job-market and sustaining interest over the longer run. In hindsight, these concerns are critical when choosing a research programme. In the end a growing interest in Marxian political economics, generated by early classes with Ruth Fincher and Michael Webber and reinforced by disciplinary debate, trumped all else. After one summer reading and re-reading Marx's *Capital* (1976 [1867]), I was absolutely hooked. So, it seemed, was everyone else: this work appeared to have legs.

With friends and colleagues working on questions of unequal exchange and uneven development, and on the transformation problem, the relationship between labour value and price, I was drawn to a small group attempting to shore up Marxian political economy from the attacks of neo-Ricardians (Steedman 1977), to formalize some of Marx's arguments and extend them in an explicitly spatial direction (Sheppard 1984; Webber 1987). Using Marxist theory of accumulation and crisis to understand industrial and regional decline also was high on the agenda (Massey and Meegan 1982; Webber 1982). This work prompted questions about the internal consistency of Marx's arguments, of the links between theory and empirical work and broader claims of method and epistemology.

A relatively simple question represented my entry into this research area. That was whether empirical information could be gathered that might support arguments for the existence of long-waves of economic activity in the Canadian economy. Rapid economic decline across much of the industrialized world, following the long post-war boom, appeared to have rekindled interest in Kondratief waves, cycles of economic activity with a claimed periodicity of approximately fifty years, and in the mechanisms that might drive such cycles (van Duijn 1983). Armed with a question, the immediate research problem now became how to look for a long-wave in economic data. What variable or variables might represent the most meaningful indices of the economy as a whole, and would information on such indices be available over a relatively long time frame? Obvious candidates were measures of technology, closely connected both to economic performance and the form of accumulation. Identification of a key variable was not approached in a theoretical vacuum. For Marxist economists, the rate of profit is the critical barometer of the capitalist economy, representing an upper bound on the rate of investment and the rate of growth of the economy.

Most political-economy studies of the rate of profit employed a very simple conception of profitability based on the costs of and the returns to production. These accounts were flawed because they failed to recognize that turnover times of capital, the period between which capital is advanced to fund production and recouped through sales, were variable. The solution to this problem involved measurement of the capital advanced to support production, of the capital sunk in owned inventory, both in complete and semi-finished products, and in estimates of the turnover times of capital (Webber and Rigby 1996). Data on the rate of profit for the Canadian manufacturing sector that spanned most of the twentieth century did indeed appear to be highly correlated with our general understanding of the long-run history of economic growth. The quantitative analysis confirmed the usefulness of the rate of profit measure developed, it supported theoretical arguments on the general form of the Kondratief cycle and it lent empirical weight to claims about the existence of long cycles of economic growth and decline.

This initial exposure to research instilled an interest in questions of technological change and on the construction and analysis of novel data sets that undergirds my work to this day. It directly inspired my subsequent dissertation research. One of the key problems in the long-wave literature was trying to explain the turning points of the cycle: what causes a phase of growth to end, and what causes the economy to snap out of recession? This substantive research problem was coupled with a growing interest in Marxian political economy, as a general platform for understanding the dynamics of accumulation in capitalism. The two interests came together when I started work on Marx's theory of the falling rate of profit. This theory combines a model of economic growth linked to the labour market and a simple model of capitalist competition based on technological rents. The argument provides an easily understood logic for building technological change into a growth model and it traces the impacts of such change throughout the economy, combining individual actions with system-wide repercussions. In short, the theory states that as the economy expands, with existing techniques of production, the demand for labour increases and will eventually drive up wages. If capitalists do nothing, they face a wage-induced squeeze on profits. If they react by adopting labour-saving technologies, the rate of profit will fall because of realization or under-consumption problems. Thus, there is no escape from periodic economic slowdowns: unfettered economic expansion is not possible in the capitalist economy.

If this argument was so straightforward, was there a need to revisit it? Did it constitute a problem suitable for another doctoral dissertation? The answer to this question has two parts. First, the logical consistency of Marx's theory of the falling rate of profit had repeatedly been challenged. Second, the theory was not so esoteric that its consistency was of no consequence. At root, the theory of crisis is the progenitor of change that underpins a significant volume of the research conducted within economic geography over the last two decades (Aglietta 1987 [1979]; Harvey 1982; Massey and Meegan 1982; Tickell and Peck 1992). Resting on these arguments are various models of technological change, models of regimes and sub–modes of capitalist production, theories of regulation and much of the writing on new forms of industrial organization, regional uneven development and labour process.

My analysis of the theory of the falling rate of profit was relatively formal, employing various mathematical tools to outline a general model of accumulation, examining how different forms of technological change impact the profitability of capitalist production as a whole. The nature of the question that I asked demanded this type of approach. The resultant model was abstract in the sense that it offered a deductive explanation of capitalist economic crisis. It was not a model that purported to explain why a specific set of firms, an industry or a region at some given time might experience a slowdown in growth. In that sense the model developed was essentialist/reductionist, the explicit aim was to identify precisely the conditions under which viable, or cost-reducing, technical changes would lead to a fall in the average rate of profit.

Tenure and the limits to capital

By the late 1980s, as I ended graduate study and moved into a faculty position, the excitement of a few years earlier was much dissipated as the future of Marxist political economy within geography, as well as outside, was in doubt. Eric Sheppard's and Trevor Barnes' excellent introduction to analytical political economy, *The Capitalist Space Economy* (1990), had just, using the authors' own words, 'sunk without trace'. This sea–change was difficult to comprehend. We still didn't have an internally consistent theory of the dynamics of the capitalist mode of production, of crises and growth. Neither did we have a very good understanding of how those dynamics were shaped by particular constellations of contingent forces giving rise to variant and restless geographies of uneven development. Nonetheless, a good number of the standard bearers of the engagement of political economy and geography were quick to take flight. Around the same time, human geographers, so we were told, were becoming increasingly dissatisfied with grand theory and with the primacy of quantitative modes of analysis. As an analytically inclined political-economic geographer, the future, and tenure, were starting to look a lot more uncertain.

These uncertainties were manifest in various ways and at different speeds. Shifts in funding opportunities, in editorial boards and the focus of disciplinary journals, falling citation levels and decreased attention to particular questions and styles of analysis are all symptoms of the restructuring of a vibrant academic community. Fortunately for me, within economic geography these changes were never absolute, as the future of the sub-field is still in question, and few of them have fully materialized. Nonetheless, a research programme focused on questions and employing approaches endorsed by smaller numbers of colleagues did not seem the best strategy for tenure. However, a radical shift in research focus prior to tenure also carries much risk. My choice, along with that of many others, was to steer a middle-course.

Since the mid-1980s, there had been general acceptance that the broad logics of a Marxian political economy worked themselves out in different ways in different times and places. But as interest in these broad logics waned, attention turned to what might be termed 'mid-level theory' and related empirical analysis. That theory was typically unconnected to more abstract structures or was loosely connected to a less well-specified 'upstream' vision of the 'economy'. The resultant work was often of a case-study variant, exploring specific patterns of industrial and regional restructuring. The 'localities' research programme in the UK is a good example of such work, though some might question whether this venture was connected to any theory at all. Much more recent work on industrial clusters, on agglomeration, innovation and knowledge spillovers is similarly 'agnostic' in terms of more abstract theory.

My work changed too. The relatively formal mathematical modelling of the dynamics of a Marxian space economy shifted towards more empirical concerns such as the long-term movement of profit rates in different industries and regions, exploring the impacts of different types of technological change on profitability, and attempts to identify the factors that influenced the convergence or divergence of rates of profit. To echo the notes above, this work was less explicitly connected to a Marxian model of the laws of motion of the capitalist economy, even though that model provided the theoretical roots for the analysis.

Moving to UCLA brought the US regional economy into sharper focus. While the USA has relatively good secondary data, much of that data had not been exploited. Thus, I perceived some competitive advantage in the construction and analysis of novel data sets. I set about developing manufacturing accounts for US states and census regions at as fine a level of industrial disaggregation as possible. An immediate problem with the US data was the lack of regional capital stock series, a critical component of measures of productivity, of technology and profitability. Production of these capital stock accounts took considerable time, but their generation sustained research in quite diverse areas, from analysis of the changing geography of manufacturing investment and the age of capital, a proxy for technology, to investigation of the strength of different forms of technological change, and to related work on spatio-temporal differences in techniques of production in US and Canadian manufacturing. This provided further evidence that equilibrium-based accounts of competition as a mechanism that simply eliminates difference were seriously flawed. Making sense of the persistence of differences in technology and measures of performance, however, called for an alternative model of competition and economic change that recognized variety in the characteristics and behaviours of individual businesses, and employed that heterogeneity to understand the evolution of technologies, industries and economies. An obvious candidate was the emerging framework of evolutionary economics. While theoretically rich, that framework did not lend itself well to empirical work, in large part because of the need for individual-level data. That immediately presented a problem, for abstract economic models, of any persuasion, were unlikely to find a welcome audience in human geography.

Around the same time, I had been searching for funds to support estimation of the regional capital stock data. That support, from the National Science Foundation (NSF), was not forthcoming, in part because an economist reviewing my proposal claimed that such data were already available. The reviewer referenced an unpublished economic database that was purported to contain records for every single manufacturing establishment in the USA. This was a revelation: here were the data that would allow close empirical examination of theoretical arguments within evolutionary economics, along with analysis of how such arguments might play out in different spatial settings.

Towards an evolutionary economic geography

It is all too obvious when we observe the economy, when we gather secondary data for firms, industries and regions, or when we survey workers or business owners, that we see an extraordinary amount of diversity. Yet, for much of the last one hundred years, the models that have dominated our understanding of the economy have been based on the concept of the representative agent. This is a descendent of Mills' homo-economicus (Persky 1995), more recently resurrected by Barnes (1987), a 'species' most starkly characterized by its homogeneity, a trait resulting from the genetic endowment of uniform rational preferences, perfect information and the ability to optimize instantaneously at zero cost. Typological approaches, that ignore variety in individual characteristics, are useful in so far as they facilitate solutions to abstract models of complex phenomena.

In so far as they eschew variety, typological approaches do not seem to offer much analytical purchase to economic geographers, especially those trying to understand the difference that space makes. An alternative is provided by those population approaches that focus on diversity in the characteristics and behaviours of the entities under study (Hannan and Freeman 1977; Metcalfe 1994). For proponents of population perspectives, system-wide change is dependent upon heterogeneity and attendant processes of selection that favour some of the variable characteristics of a population over others.

Evolutionary models provide perhaps the most well-known population-based approach to understanding economic dynamics. Evolutionary accounts of economic dynamics rest squarely upon heterogeneity in the characteristics and behavioural routines of firms and other economic agents (Nelson 1995; Nelson and Winter 1982; Saviotti and Metcalfe 1991). According to these accounts, production within the capitalist economy is controlled largely by profit-seeking firms. No firm is guaranteed profit, for the market is a chaotic arena where prices for inputs and outputs cannot be determined a priori. Some firms attempt to manage this uncertainty by controlling the market. However, the majority of firms can only control the manner in which they transform inputs into output, seeking to achieve a competitive advantage by increasing the efficiency of their production. For most firms, efficiency is unknown until they enter the market. In this competitive environment firms are compelled to innovate, to search for new products and develop new markets, to experiment with new sources of inputs, new processes of production and organizational routines, sure only in the knowledge that others are doing the same. It is this search for efficiency, under conditions of uncertainty, that generates heterogeneity in the characteristics and behaviours of individual firms.

At the same time, competition also destroys heterogeneity through processes of selection (differential plant growth), imitation and firm exit (Foster and Metcalfe 2001). In the market, the differential allocation of profit across firms tends to favour (select) individual routines, specific techniques of production, products and even entire industries (Baldwin 1995; Dunne et al. 1988). Within the selection environment, individual firms tend to occupy relatively distinct locations. They are unable to mutate instantaneously and at zero cost, constrained as they are by various forms of organizational inertia and knowledge sets that are bounded in large part through experience and a limited set of interactions. While inertia holds variety constant long enough for selection to operate, familiarity with established routines and the costs of their abandonment constrain patterns of search for new routines and their adoption. Path dependence, irreversibility and 'lock-in' thus give rise to firm, and in aggregate, industry- and region-specific patterns of technological change and behaviour (Arthur 1994; Storper 1992).

While the arguments from evolutionary theory are compelling, especially in terms of the dynamics of competition, the central role ascribed to heterogeneity has rarely been scrutinized empirically. My work with Jürgen

Essletzbichler explores whether or not heterogeneity in plant characteristics, particularly technology, exhibits a significant spatial dimension, and how geographically localized patterns of heterogeneity shape the performance and evolution of different regional economies. In fact, regional differences in techniques of production are statistically significant and they tend to persist over time; moreover, trajectories of technological change and components of productivity growth depend on region-specific patterns of heterogeneity in techniques of production (Rigby and Essletzbichler 1997, 2006). Heterogeneity seems to be a permanent rather than a transient feature of the economy. It also exhibits a strong geographical imprint, manifest in the regional accumulations of knowledge, organizational and behavioural routines and firm characteristics.

The empirical dimensions of this analysis are crucially dependent on plant-level data. It turns out that such data are available for a large number of countries, but access, because of the confidential nature of individual-level information, is extremely difficult. Access to these data nominally rests upon approval of a research proposal along with payment to cover the costs of data preparation and support. Legal and administrative hurdles place heavy burdens on those seeking access. Nonetheless, for geographers and others interested in spatial-economic dynamics, these data represent a treasure trove of possibilities. Of course, primary data collection through plant surveys is an alternative, though the cost of such work, spanning industries and regions across the USA, is prohibitive.

Conclusion

The path of my research has been strongly conditioned by the academic environment in which I am embedded. That environment changes over time: it evolves as some questions and approaches become more highly valued than others. This movement is not necessarily a bad thing, indeed, such shifts within an academic community are surely evidence of a certain vibrancy. However, such movements are not always positive and certainly the recent evolution of economic geography is difficult to understand. At a time when questions of the difference that space makes appear to be increasingly relevant, economic geographers have seemingly lost their voice. Economic geographers are not prominent figures in cross-disciplinary debates on a host of key issues such as globalization and its impacts; on emerging patterns of urban or regional competition and economic change; on industrial evolution; on labour market dynamics; on agglomeration, learning, technological change and growth; or on broader questions of development.

How do we explain this? Some would undoubtedly proclaim these sorts of questions as far too restrictive a way of conceptualizing the domain of economic geography and hence the gaze of many economic geographers has rightly settled elsewhere. While I would agree to some extent, it does seem to me that a broad focus on the economic condition of people and places should be key. And, if pushed further, I would argue that the declining relevance of economic geography has much more to do with the current poverty of theory in the sub-field and a peculiar relativism of method that seems ill-adapted to tackling substantive questions. We all proclaim the difference that space makes. However, beyond simple appeals to differences in the character of particular places, how do we show why geography matters and how do we convince others? In large part, we need to develop or extend bodies of theory that show how a concern with space provides novel insights into the functioning of different parts of the economy (broadly understood). Right now, a good deal of such action appears to be taking place within the rapidly growing sub-field of geographical economics. Dismissing all such work as non-geographical and hence irrelevant does not seem to be very productive. Surely, we need to provide the skills that will allow future economic geographers either to challenge the geographical economists on their own turf (that is, after all, ours), or to develop more sophisticated alternatives that have similar scope. And that means training new economic geographers in economic theory of a form that is logically consistent, capable of accommodating insights from geography, and that can be evaluated in some meaningful way.

15 BEYOND CLOSE DIALOGUE: ECONOMIC GEOGRAPHY AS IF IT MATTERS

Gordon L. Clark

Any debate about social science method must be sensitive to the social context in which discussion takes place. So, for example, in the 1970s and the 1980s debate about method and the goals of social science was heavily influenced by the social turmoil occasioned by the Vietnam war. Simplistic notions of social science as a science without political and ethical commitments were overturned by a new generation deeply suspicious of the claims of neutrality and higher purpose made by their teachers and supervisors. Twenty-five years on, the debate about the purpose of theory-building is informed by similarly contentious issues such as neo-liberalism and the putative imperialism found, by some, in the 'war on terrorism' (Harvey 2003).

One of the distinctive features of economic geography as practised by geographers is its commitment to understanding the world as lived – an empirical project – rather than the world as theorized. This is perhaps a bold assertion, one that may be contested by competing claims and counter-examples. But in a general sense, comparing economic geography as practised by geographers with economic geography as practised by economists, we seem to be a discipline more inspired by the world around us than by cathedrals of theory. Why this is the case is hard to fathom. It may have something to do with the type of people attracted to the discipline, and it may have something to do with the fact that deductive thinking claims a higher rate of return in economics than in geography.

Over the past thirty years, building a story out of local detail and then drawing implications for understanding the rest of the world has been very important in geography. It seems entirely normal to begin with an observed problem and then move to analyze it in terms of its shape, scale and significance, bringing to bear the theory at hand. This tells us something about our intellectual impulse and how we tend to start off with an observed problem even if we do not conclude with it. For it is important to acknowledge that even if we begin empirically we almost always end the story theoretically. This means I am rather sceptical of the theory-first school of thought: there is a paradox deeply embedded in the practice of social science; there seems to be insufficient theory to make comprehensive sense of the world and yet there seem to be many rival theories in relation to what we know about the world (Clark 1998). Being sceptical about theory, however, means that we must be rigorous about how we use the empirical world to build explanation and understanding.

This chapter reflects my unease about the apparent open-endedness of close dialogue and the pressing need to formulate more precise tests of attitude, opinion, and ultimately action. In the next section I look at the purpose of social science in general, and the conventions that underpin economic geography in particular. I make the case for a constructive social science, one that emphasizes what we understand about the nature and structure of the world. I subsequently look back on the idea of close dialogue and show how it has evolved and has been strengthened through a variety of more formal qualitative and quantitative techniques. This takes us from tests of consensus, to Q-methodology, and to psychometric testing. In the penultimate section, I look again at the crucial intellectual problems to be solved in an economic geography capable of tackling the biggest problems of humanity. Finally, I look at the emerging research programme about the local and the global.

Purpose of social science

As an institution and as a social practice, social science should add to our knowledge and understanding of society (here and there). It should combine the past and the present in ways that are deliberately analytical and evaluative. As described, this is little different from the mission statement of any science, natural or otherwise. The natural sciences could claim to be similarly committed to the well-being of society; but in many respects their commitment to the well-being of society is more abstract and less tangible than the commitment of the social sciences. The natural sciences are less connected than the social sciences to the formulation of contemporary public policy, given their very different spatial and temporal time horizons. Social science is a *constructive* enterprise, concerned (in part) about social justice as an ideal.

Social science should also add to knowledge and understanding of society in ways that can be *communicated* to others. Unlike the natural sciences, however, there is no universal language such as mathematics that can join us together within and across societies characterized by different cultures and languages. I recognize, of course, that this is sometimes expressed as a dream for the social sciences. But such dreams are fantastic. Part of my scepticism on this count can be attributed to scepticism of formal symbolic reasoning in circumstances where concepts are socially constructed and their meaning negotiated through social institutions and relationships. I am also sceptical of the power of symbolic reasoning in that social life is quite variable over time and space; the empirical world is sufficiently different from time-to-time and place-to-place to allow for heterogeneous understanding of common economic and social processes (see Strange 1997).

More controversially, I would contend social science is innately *democratic* in that the purpose of adding to knowledge and understanding of society (here and there) is to empower others not so privileged by academic position and association. This ethos carries with it two implications. First, social science should produce in a systematic fashion knowledge and understanding that others are able to appreciate and commonly accept as plausible. Without standards of value in the production of knowledge and understanding all we may have, at the end of the day, are spectacular moments of self-aggrandizement that have little to do with the experience of those who are the objects of social science research. Second, if social science is constructive in the sense of building knowledge through which others may change society, then the knowledge produced must be clear as to the most important variables and their causal relationships. In this respect, the true test of the power of social science knowledge is not so much its predictive ability but its capacity to produce maps of causality that allow others to assess alternatives for intervention.

If social science is a constructive enterprise, it should also be a *critical* enterprise. We must be critical of what we know, and how we know it. Furthermore, we should recognize that knowledge changes as societies evolve in time and space. Knowledge may improve in the sense that we get better at being more systematic about important variables and their causal relationships. But given that society evolves over time and space, knowledge of society has to be continuously regenerated or reinvigorated to capture that evolution. As society changes, the knowledge and understanding of one era, such as the 1950s, becomes difficult to transplant into another era, such as the 1990s and the first decade of the twenty-first century. We may be able to take forward institutions such as markets and economic systems, but those are, more often than not, interpreted according to different standpoints in time and space.

I am mindful that this mandate may horrify those in economic geography at the intersection between culture, language, and economic process. But there is no reason to think that this mandate for economic geography as a social science would exclude that kind of research programme. Quite the contrary. I would suggest that culture and knowledge, social identity and location, economic structure and process etc. are intertwined (Thrift 2005). This is why I emphasize knowledge and understanding – code words for structure and interpretation (Clark 2001). But I am impatient with arguments to the effect that economic geography can be content with simply documenting one element or one set of elements as if the expression of difference in one setting is sufficient to legitimize the whole enterprise. The purpose of social science is to be found in its commitment to constructing systematic maps, if you like, of social life.

Close dialogue reprised

One reason for the immediate take-up of close dialogue was its reference to contemporary practice in economic geography (Clark 1998). Similarly, the paper linked to related arguments made by others, especially in fields of inquiry such as gender studies. The argument gave respectability to research practices that had developed over the 1980s and 1990s when large-scale quantitative models were eschewed in favour of case studies and interviews. It is arguable that successive generations of economic geographers, not trained in conventional quantitative methods but nevertheless driven by inspiration and ambition, sought alternative means of telling the story about the transformation of corporate capitalism. Indeed, it is arguable that only those stories could have brought to the fore the penetration of financial institutions and markets into each and every community (Clark 2006).

Economists interested in economic geography, when made aware of this rather different research practice and the priority attributed to inductive reasoning as opposed to deductive reasoning, were less than complimentary (Krugman 1996). But there are certain ironies in their reaction; after all, some of the most important economists interested in economic geography were themselves cast as renegades in a discipline noted for convention and its commitment to the axioms of neoclassical theory. The postulated existence of increasing returns to scale, for example, was used to overturn conventional assumptions about the inevitability of a smooth landscape of income and employment opportunities. Yet the use of close dialogue to interrogate conventional assumptions about market efficiency and the space-time dynamics of capital accumulation was deemed beyond the pale.

For many geographers, the use of case studies to interrogate theory has a genuine claim of legitimacy. It would be very odd indeed if data were selected to be consistent with theory such that theory was always reinforced by a positive message as seems to have been the case in finance with respect to the efficient markets hypothesis (see Lo 2004). Lack of sensitivity to counter-examples simply encourages the building of empty castles in the sky. More particularly, counter-examples if explored in their full dimensions with respect to their points of departure as well as their points of agreement with received wisdom, carry with them important lessons for the scope of received wisdom and its vulnerability. The problem comes when stories told are disconnected from their points of reference in received theory. In these situations, unstructured stories may wash away the theoretical impulse.

As suggested above, I am by both impulse and practice sceptical of theory if it is an iron-clad set of principles that allows us to deduce the world as lived by ourselves and others. Given our interest in spatial differentiation, the proper theory of the world is a theory that must account for both commonalities between places and, if not the precise detail of difference, at least the existence of difference. Theories of regional growth and development, for example, appear too abstract to cope with the diversity of what we have at hand and we seem unable to differentiate between the theories that work and the theories that do not work when trying to explain the world. This is not an argument for rejecting theory-qua-theory. Rather, inspired by David Hume among others, I tend to work backwards and forwards between theory and the empirical world in a reflexive manner seeking *rapprochement* rather than climbing the ladder of theory to the peak of perfection.

I would also suggest that the process of mutual engagement has deep roots in cognitive psychology, being consistent with how people actually conceptualize the world and make decisions. Those who advocate theory first do so knowingly or not in the shadows cast by deductive reasoning and logic. Of course, academics may argue the case for the *proper* way of knowing, even if most people do not do so. There comes a point, however, where such idealism becomes elitism and a blueprint for knowledge-making that is profoundly undemocratic. Some academics are dismayed by this kind of practice, believing that referencing individual cases is hardly appropriate in systematic policy-making. On the other hand, the significance of case studies in political discourse tells us something about how people actually think and how people use evidence at the intersection between received wisdom and social aspirations.

Calibrating close dialogue

As practised, close dialogue is a mode of conversation and interviewing that sets received opinions against informed expectations. Those expectations are a combination of experience *and* theoretical predispositions, recognizing that a single interview is unlikely to overturn either on the basis of one conversation. Unlike most ethnographic modes of research, however, close dialogue is less about giving 'voice' to others' experience and values than it is about collecting the raw material for empirical analysis and theoretical synthesis. Close dialogue is structured in another sense; it is often guided by a set of preconceived questions rather than being allowed to wander question-by-question following the respondent.

For these reasons, I rarely identify respondents or quote their opinions directly in the text. This would be both misleading as to the intention behind close dialogue and unfair in the sense that those identified do not have the opportunity to defend themselves or their opinions in relation to the engagement with theory that characterizes publication based upon close dialogue. I have argued elsewhere, in fact, that to explicitly identify those interviewed in this process of accommodation between the empirical world and the theoretical world may be to leave them responsible for the research process as much as researchers' views derived from the research project (Clark 2003). As close dialogue is about economic agents and their circumstances, it is also about the researcher and his/her aspirations, wherein the latter inevitably dominates the former. There remain, however, a number of shortcomings associated with close dialogue. Most importantly, there are two questions at hand: how can we build generality in a systematic way? And how can generality be validated outside a presumption of trust in the researcher's motives?

Close dialogue is normally done on the basis of one-to-one interviews. Another way of approaching the dialogue process is to collect respondents together in a conference setting or to invite them to join a research colloquium that has, as its explicit goal, the development of knowledge and understanding. Industry-based research programmes often use the former whereas, increasingly, academic research programmes utilize the latter. In these settings, respondents are brought together and seated either at random or with purpose (e.g., representation) in groups of as many as ten people with electronic response consuls that allow participants to indicate their views in an immediate fashion. This technique can calibrate responses: by posing questions, eliciting debate, recording individual responses, summarizing responses, and reporting responses to participants for further recorded response. Researchers can both gauge the scope of response to questions and then gauge whether there is convergence to a consensus view among those made aware of what one another thinks.

Bringing together large groups of potential respondents is a challenging undertaking. It is most effective when done in combination with sponsoring organizations that can claim the attention of those identified by virtue of their community, professional or business commitments. Another way of proceeding is through Q-methodology. Basically, this technique utilizes the one-on-one interview process but does so after first identifying crucial words and concepts related to both theoretical expectations and the world in which respondents live and work. These cues are used to focus the interview process and deliberately test respondents' knowledge and understanding of the particular issues that these words and concepts mean to represent. After an initial overview, researchers normally introduce cue cards and ask for recognition and response that can be summarized and coded for subsequent statistical analysis. See, for example, recent papers by Babcock-Lumish (2005) and Jiménez (2005).

Q-methodology carries with it the prospect of 'counting' multiple interpretations of common words and concepts and can be thought to be as much about hermeneutics as it is about a common language of knowledge. Moreover, it assigns priority to the knowledge and understanding of respondents by virtue of their location in time and space. Remarkably, Babcock-Lumish (2005) was able to show that there is a systematic geography to knowledge and understanding of the venture capital process (comparing London and Boston) that might otherwise be missed by researchers using close dialogue to synthesize responses against their own theoretical predispositions. As such, Q-methodology combines generality with validity in ways that allow researchers to look back at their own predispositions and forward to the heterogeneity of experience.

Most importantly, psychometric testing can explicitly link theoretical expectations or hypotheses regarding the logic of decision making to the types of problem respondents must solve on a day-to-day basis. This analysis relies, in the first instance, on researchers' knowledge of respondents' experience as well as knowledge of the theoretical issues they wish to evaluate. In this sense, close dialogue is the first step in the process of psychometric testing and a necessary ingredient in framing theory-questions so that their relevance is clear to respondents.

Inevitably, to make this kind of testing regime work we need skills organized in research teams combining, for example, knowledge of the world of respondents as well as the complement of skills necessary to translate academic concepts into sets of puzzles that are relevant to respondents *and* capable of being analyzed through statistical routines. Equally important, however, is the issue of gaining the co-operation of respondents who may otherwise prefer a less rigorous and invasive form of question-and-answer interviewing. Nevertheless, if realized with sufficient attention to each and every stage in the testing regime, the results may have far-reaching implications for the social sciences, including economic geography.

To illustrate, we have been interested in the temporal and spatial frames used by respondents to judge investment options (see Caerlewy-Smith et al. 2006; and Clark et al. 2006a, 2007). Are they short-term oriented or long-term oriented? Do they prefer close-by opportunities, and how do they value distant opportunities? Most importantly, are the skills necessary to solve financial puzzles generic (like any other puzzle) or specific to the context at hand (their roles and responsibilities)? Are the experience, gender, age, and education of respondents important? Answers to these questions may have enormous implications for the management of financial assets and the efficiency of global financial markets. To undertake such a project requires a set of skills and experience rarely found in academic departments, let alone in an individual researcher.

Research frontiers in economic geography

Many will be aware of the enormous interest in issues of individual decision-making occasioned by the path-breaking work of economists, decision scientists, and neuropsychologists (see Kahneman and Tversky 1979). Being challenged is the very idea that people are rational in the sense that they act deliberately in accordance with universal rules of logic and reason. In part, research on these issues has focused upon the cognitive skills and capacities of individuals seeking evidence for systematic patterns of so-called 'non-rational' behaviour or at least that behaviour which goes against the precepts of logic and reason. This research agenda is based upon quite specific populations in quite sterile environments. It is remarkable that the base population for much of this research has been undergraduate and MBA students in elite US and European universities, wherein they are required to deal with generic decision-making puzzles in which they have little or no stake. Left out of analysis, more often than not, is the relevance of social identity and cultural location.

Also left out of this research programme is assessment of the role of context in decision-making, and whether it makes a significant difference (or otherwise) to decision-making holding constant cognitive skills and capacities (but see Bertrand et al. 2005; and Clark et al. 2006b). This is acknowledged to be an important gap in the research programme, but one that related disciplines have found hard to bridge or even conceptualize in terms of the geographical scale of social and economic processes. The theory-first logic of economics and political science, for example, has left those disciplines without sufficient experience to deal with the intersection of cognition and context. But, of course, this is not the case for economic geography; we often argue that context matters although we have done so by intuition and illustration rather than experimental procedure.

Making good on our comparative advantage, in this respect, is very important for the future of economic geography. Our experience with close dialogue, as well as our ability to handle the complex interplay between culture, institutions, and the nature of every-day decision-making through case studies, is rare in the social sciences. If we are able to formalize close dialogue through more rigorous experimental procedures, the discipline may have much to offer in articulating the interplay between context and decision-making. The trick, however, is to translate our experience into a systematic research programme that is theoretical as well as empirical.

At the other end of the spectrum, there is enormous interest in the nature and functioning of global financial markets. For many years, research in finance was dominated by the efficient markets hypothesis so that it ruled out so-called irrational behaviour and the circumstances of different places and different times. All was well for about two decades. However, as the TMT (Technology, Media and Communications) bubble gathered momentum in the late 1990s and came crashing down around 2000/2001, it took with it confidence in the language of finance. Hard questions have been asked about the empirical plausibility of the hitherto unchallenged assumptions underpinning the efficient market hypothesis. Even those with a vested interest in the hypothesis had been forced to concede that market behaviour is different in scope from what was expected (Fama and French 2005).

One consequence has been an enormous flurry of research on three threads of interest. First, there has been a concerted research programme developed around the existence and claimed persistence of behavioural anomalies in financial markets. It is apparent that heterogeneous expectations and behaviour co-exist at any point in time and space. In the absence of a comprehensive winner-take-all reward system, heterogeneity may persist within and between markets. Second, it is apparent that many measures of value in the market have proved to be poor predictors of winning and losing stocks. This has brought forth a wide variety of alternative measures of performance, all with their own reliability and tests of relative value. The market for valuation has exploded; competing ideas, principles, and methods of empirical analysis have enormous implications for issues such as environmental sustainability. Third, there has been renewed interest in the institutions and management of inter-market arbitrage and the fact that markets vary considerably in terms of their social structures and behavioural patterns.

The discipline of geography could be at the very centre of research on global financial markets. But to enter this market of ideas requires research which takes place at two levels: in terms of observed patterns of market trading, market prices, and market arbitrage and in terms of the apparent behavioural imperatives underpinning distinctive geographical patterns of trade and counter-trade (Clark and Wójcik 2007). Enormous databases require highly sophisticated analytical routines as well as large-scale computing power to process the data. In addition, knowledge is needed of particular places and times, including reference to cultural expectations, social relationships, and economic power reflected, for example, in the relative status of insiders as opposed to outsiders in some countries' financial markets.

By this logic, close dialogue should be an important partner in such research programmes. This is possible by virtue of the fact that the core theoretical precepts of the theory of finance are in a shambles and that heterogeneous market expectations and performance have become the points of reference in analysis. With the world of finance turned upside-down and a common perception that theory-first is a losing proposition, knowledge of the empirical world is the source of making a profit. Since there is such an intimate relationship between finance academics and finance practitioners, the former cannot afford to ignore the latter. Consequently, knowledge about how and why markets are internally differentiated and knowledge about the nature and persistence of differences between markets are at a premium. To go further, we must find a way of sustaining partnerships with other disciplines so that we are party to building the theoretical logic that will ultimately replace the efficient markets hypothesis.

Conclusions

I stand by close dialogue as a way of understanding the world even if it is different from the kinds of method advocated in related disciplines. Most importantly, it should be recognized that close dialogue resonates with long-standing traditions in the discipline of geography, concerned, as it has been for centuries, with the circumstances of specific times and places. In this chapter, I suggest that there are good reasons to value this kind of thinking in the face of the claims of virtue made on behalf of deductive reasoning. In part, my claim is based upon the qualitative insights to be derived from close dialogue. But as well, it can be observed that it is a mode of analysis widely shared among the entire human race and should not be idly discounted in the face of theory-first idealism.

It was suggested that economic geography should be engaged with social science. In doing so, economic geography should be committed to adding to the knowledge and understanding of society (here and there) in ways that contribute to the welfare of the whole. Economic geography should be a constructive enterprise being about building the knowledge and understanding necessary for the progressive enrichment of human life. It also should be constructive in another way, being deliberately about developing theoretical perspectives on 'society' in all its manifestations. This does not imply that economic geography is any less critical or argumentative than other fields of research. By its very nature, close dialogue is a valuable instrument for interrogating and criticizing established theory in the interests of building something better for society. Notice, I am impatient with those who suggest that economic geography is just about documenting difference as if the existence of difference in its own right is so valuable to justify the field.

I have also argued for an economic geography that can integrate the local with the global, working from behaviour through to observed spatial patterns in markets and institutions. This is not just a mandate for research in economic geography; it is a mandate that could be shared across the social sciences and in disciplines such as finance and economics that hitherto avoided the issue through simple-minded assumptions about the necessity of rational decision-making and the efficiency of markets. Given the skills of economic geographers, it seems inevitable that alliances and partnerships should be formed with others to tackle this mandate. In any event, I have argued that close dialogue will remain a crucial way of knowing about the world and should not be seen, whatever the criticism, to be a less valuable or a somehow unjustifiable way of knowing (compared to the alternatives).

Of course, part of our commitment to understanding society (here and there) must be about globalization. Close dialogue has a vital role to play in picking apart and disputing what has become a totalizing discourse on globalization. Just as in other times and places where there has been one dominant way of thinking about the world, those who claim globalization is the organizing logic of the present and the future do so by stripping out of the analysis the circumstances and behaviour of people in their own places and times. If we are going to say something sensible about the local and the global, we have to say something about human behaviour and the effect of different scales of life on behaviour. And we have to do so in ways that allow us to go from case studies and examples to a more general theory of behaviour itself. In other words, we must take context more seriously just as we have to take 'cognition' more seriously; we have been so fearful of the ecological fallacy that the ways in which circumstances structure thinking and behaviour have been left out of the picture. By contrast, some economists practising economic geography have sought to reintroduce this theme even if many geographers have been alarmed at some of the more obvious implications.

If we take seriously the interaction between cognition, context, and behaviour, we must also think about what is 'behaviour'. One of the costs associated with the hegemony of the rational actor model has been the denial of the full scope of behaviour, including its less savoury aspects, whether that be violence, corruption, or exploitation of others. People are moral and not so moral beings and their 'behaviour' must be catalogued and explained by reference to that variety. Unfortunately, for more than a generation 'behaviour' has been stripped bare of its scope and context.

Acknowledgements

This chapter was inspired by collaboration with Dariusz Wójcik, Terry Babcock-Lumish, Emiko Caerlewy-Smith, Orlando Jiménez, Kendra Strauss, and Adam Tickell. I benefited from contributing to a seminar on methodology at a meeting of Scandinavian doctoral students at Götenborg University (Sweden) hosted by Anders Malmberg and Bjorn Asheim. I would also record my appreciation for the comments and advice received from Trevor Barnes, Harald Bathelt, Linda McDowell, and Kendra Strauss. None of the above should be held responsible for any errors, omissions, or opinions contained herein.

16 ECONOMIC GEOGRAPHY, BY THE NUMBERS?

Paul Plummer

Quantification in context

A common perception among contemporary economic geographers is that quantification is either dead or dying: an approach that is a historical curiosity, a remnant of a misguided attempt to replicate the methodological norms and practices of science (Johnston et al. 2003; Yeung 2003). As a social construction, an 'ideal' set of methodological norms and practices that we label 'quantification' is supposed to be derived from a failed and discredited social ontology and epistemology, grounded, as it is, in the modernist project of positivism/empiricism (Johnston 2006). Following the publication of Andrew Sayer's (1984) highly influential *Method in Social Science*, someone practising quantitative economic geography is supposed to engage in 'extensive' research, employing large-scale survey and statistical analysis to search for regularities among map patterns, thereby producing descriptive accounts lacking in explanatory power. In contrast, the preferred suite of qualitative methodologies are characterized as 'intensive' case studies, examining individual agents and their understandings in their causal contexts in an attempt to uncover the processes generating observed outcomes. I find this characterization of competing research designs deeply frustrating, bearing little relation to the way in which I, and others, engage in quantitative economic geography. This stereotype is promulgated through an unsustainable dualism between quantitative and qualitative ways of knowing (Plummer and Sheppard 2001). Equally troubling is the fact that this dualism appears to be widening the gulf that separates competing economic geographies, creating unnecessary and mutual antagonisms (Plummer and Sheppard 2006).

Published in the 1980s at a time of intense intellectual ferment about the foundation of economic geography, Sayer's critical-realist manifesto represents a watershed in debates about methodology in economic geography (Maki et al. 2004). Looked at through the eyes of many contemporary economic geographers, critical realism may have been superseded by variants on a post-modern methodology, nonetheless the dualism between 'extensive' and 'intensive' research strategies has remained hugely influential in shaping contemporary discourse and research training, at least in the UK.

At best, quantification seems to be tolerated in so far as it is capable of providing background information as a prelude to the real, and preferred, business of 'intensive' research. At worst, faced with the apparent inadequacies of superficial 'extensive' methodologies, many economic geographers have simply abandoned quantification completely, in favour of a suite of supposedly 'deeper' causal explanations provided by qualitative methodologies (Massey and Meegan 1985b; Clark 1998). It is but a short step from critique based upon such stereotypes to outright dismissal and institutional marginalization. The construction of quantitative geography and geographers as somehow not on the 'cutting edge' of research in economic geography is likely to produce an acute sense of paranoia among the 'old' and an increasing sense of insecurity among the 'young'. Who would want to participate in a failed enterprise, something that we used to do, and that is of limited relevance in answering questions about those aspects of social reality that the 'great and good' think call for an explanation?

Where to begin? Reports of the death of quantitative economic geography are much exaggerated. Certainly, there are still plenty of economic geographers who continue to practice quantitative methodologies. With the recent resurgence of geographical economics, some have once again sought to appeal to the strictures of 'scientific' methodology to justify their research practices (Brakman et al. 2001). However, it is possible to reject positivist/empiricist philosophies without abandoning quantification and empirical modelling in economic geography (Rigby, Kwan, both in this volume). Given the onslaught on the rigour and relevance of quantitative economic geography over the past two decades, what is clear is that there is a need to defend quantitative research practices in economic geography (see Barnes 1996). My view is that there is something worth saving: the positivist 'ideal' may be neither attainable nor desirable, nonetheless quantitative methodologies can provide us with critical understanding and explanation of social reality which can both inform policy debate and allow us to evaluate empirically our theoretical claims.

Rather than rehearse these methodological debates here, I want to move the discourse forward by engaging with the theory and practice of quantitative economic geography in the context of its own application. In particular, I want to foreground the status of 'truth' claims that are associated with employing empirical models to bridge the gap between 'theory' and 'data'. I focus on the question: what, if anything, is the value added by quantitative economic geography? What we expect to achieve in practice raises some non-trivial epistemological questions regarding the potentials and limitations of empirical modelling. Should we aim for 'true' models of the mechanisms and processes operating in the geographical world? Alternatively, are we limited to building empirically adequate models that are, perhaps, approximately accurate and precise representations of the empirical data? More broadly, should we judge empirical models by their correspondence with the 'facts', their coherence with accepted belief systems, or their usefulness in solving problems?

These are fundamental questions that deserve attention by all economic geographers regardless of their methodological commitments. I will confront these questions directly by critically examining my own methodological practices through the lens of collaborative research that I have been conducting into the dynamics of uneven development across Australian local labour markets. This entails the use of spatial econometric models and methods to empirically test popular 'institutional' theories of local economic growth (Plummer and Taylor 2001a, 2001b). Typically, such theories are constructed and empirically validated using 'intensive' research designs and qualitative methodologies, rather than empirical modelling methodologies. This research puts into sharp focus questions of ontological commitment and epistemological standards of success when employing empirical models in economic geography.

Quantification in theory

In principle, empirical modelling combines the prior belief of the investigator, substantive theories that are considered pertinent to the problem at hand, empirical data derived from observation, and a logic of inference intended to bridge the gap between 'theory' and 'observation' (King et al. 1994). In any plausible modelling scenario, these tensions are translated into multiple potentially conflicting model design criteria that range from simplicity and goodness-of-fit through to explanatory power in understanding the evolving economic landscape and success at predicting future landscapes. Regardless of our model design criteria, at each stage in the modelling process methodological choices must be made about how to proceed in order to achieve our stated objectives (Kennedy 2002).

A fundamental source of disagreement *within* empirical modelling methodologies is the relative importance assigned to 'theory' and 'data' in drawing inferences about the operation of actual social processes. Those who advocate data-driven methodologies prioritize data-mining techniques that can be used inductively to search for empirical regularities and seek generalizations between and within map patterns (Openshaw 1997). Such methodologies tend to confirm critical realism's stereotypical representation of 'extensive' research designs as a largely descriptive enterprise. I share this scepticism regarding the validity of such data-driven methodologies, but for different reasons. Data-mining techniques may appear to be successful at discovering empirical regularities without the need to impose a prioritheoretical claims about the causal mechanisms that might have generated such regularities. However, they tend to downplay both the degree to which empirical regularities are constructed in light of existing substantive and measurement theories, and the degree to which such derived empirical regularities have the potential to reflect the search paths that were selected, rather than the actual process that generated the observed data.

A healthy scepticism regarding data-mining methodologies, combined with a careful reading of recent literature in the philosophy of science and econometric methodology, led me in the opposite direction when trying to understand the dynamics of local growth in the Australian context. Standard econometric 'textbook' methodologies prescribe an essentially theory-driven approach to model specification and testing that follows the logic of a broadly defined 'deductivist' methodology (Lawson 1997). Specifically, observation claims are deduced from some theory, tested against a suitably defined data set, and subsequently judged to be confirmed or falsified depending on the nature of the observation claim and whether that claim turns out to be true or false. Operationally, Maddala (2001) suggests that theory-driven methodologies translate into a four-step sequential model design and validation process: (1) *formulation* of an empirically testable model specification based upon economic theory and prior knowledge; (2) *estimation* of the (set of) empirical model(s) using the appropriate statistical techniques and suitably derived data; (3) *testing* hypotheses suggested by the theory(ies); and (4) *prediction* and *policy analysis*. This methodological strategy is replicated in standard econometric software designs (e.g., PcGive or STATA), which proceed from model specification and selection of an appropriate estimation method through to post-estimation diagnostics, including hypothesis testing.

Given appropriate 'data' and substantive 'theory', the standard econometrics literature provides us with clearly defined algorithms that tell us how to select 'good' estimators, how to empirically test model specifications against a set of alternatives, and how to test the statistical assumption underpinning our chosen estimator using mis-specification tests and graphical diagnostics. While textbooks are strong on providing such codified knowledge, they are relatively weak on suggesting how we might develop our 'know-how', or tacit knowledge, when designing and evaluating empirical models (Magnus and Morgan 1999) and they rarely help in addressing key methodological questions about what constitutes appropriate model design criteria, how we might choose between models that satisfy these criteria, or how to formulate empirically testable models from available theories and prior information.

The conventional emphasis on codified knowledge is unfortunate since developing tacit knowledge would seem fundamental to developing sound empirical methodologies and training future generations of quantitative economic geographers. We should be explicit about our decision-making regarding which theories are relevant to the problem we are addressing, how we might translate those theories into empirically estimable models, what data are appropriate to test those theories, and how we should interpret our empirical results. In my experience, developing tacit knowledge is integral to effective empirical modelling, and essential when things go wrong, as they invariably do in practice.

Quantification in practice

I now turn to an example of my own empirical research on the nature and determinants of local economic performance in Australia. This is part of a broader collaborative research project that I am undertaking with Mike Taylor that uses multi-methods research strategies to critically engage with popular 'institutionalist' approaches to understanding the processes driving local economies. Our research project is informed by the perspective that there is a lack of systematic empirical research to support the increasingly complex theoretical arguments that are being constructed to account for the significance and persistence of local uneven development (Hudson 1994). Set against this broader set of objectives, there are lessons to be learned from reflecting on how we conducted our modelling exercise, the implications of the methodology chosen for how we might interpret our empirical results, and how we might do better in the future.

The task of translating a set of 'institutional' theories into a suite of empirical models from which it is possible to deduce a testable set of observational claims turned out to be very difficult. This is hardly surprising, these theories are typically based upon vague and ambiguous constructs, such as 'human capital', 'social capital', 'embeddedness' and 'institutional thickness' (Durlauf 2002; Plummer and Taylor 2001a). One option is to conclude that theoretical constructs lacking in precision and clarity are so 'chaotic' that they should be abandoned and that we should give up the whole enterprise as a bad job. Better, more clearly specified, theories are certainly to be welcomed, and there were times when this seemed like a preferred strategy. Nonetheless, if our theoretical claims are to be more than just clever words, at some point we need to confront theory with evidence.

We begin by identifying a suite of variables that could be rationalized as plausible measures for the hypothesized drivers of local economic growth. In practice, selecting appropriate measures is messy, open-ended, and subject to iterative processes which, hopefully, converge on a mutually agreeable set of definitions. Implicitly, translating between theory construct and empirical measure is a two-stage process, involving methodological choices at each stage. First, the key drivers of local economic growth must be identified for each of the competing theories of local economic growth: a formidable challenge. The diverse literature on 'institutional' theories of local economics growth reveals a multiplicity of competing, ambiguous, and sometimes contradictory definitions of key variables that in some cases seem to border on the tautological. Sometimes, theories appear to have been developed in a vacuum, with no intent on empirical evaluation, econometrically or otherwise. Second, for each theory these identified drivers needed to be translated into empirical measures. At this stage, we made a key methodological decision to map single empirical measures into each theory construct: human capital is measured in terms of higher education, institutional thickness as the effective rate of protection of industry in a local economy (see Plummer and Taylor 2003: Table 1). We would not claim that our translation between theoretical construct and empirical measure is either unique or incorrigible. We could have made different decisions using different variables, or combinations of variables, to define key theory constructs. This could have resulted in different model specifications and, potentially, different results when these 'theories' are confronted with 'data'. Given the uncertainty attached to this operationalization, we view our conceptualization and representation of 'institutional' theories as a step forward in explaining the dynamics of local economics, subject to revision and possible rejection in the light of further theoretical and empirical evidence, including critiques of our empirical modelling exercise.

Identifying potential drivers of local economic growth is one thing, specifying how those variables should be related to one another is quite another. Problematically, the suite of 'institutional' theories offered little a priori guidance about the nature of the hypothesized relationship between these drivers and local economic performance. In particular, the theories are singularly unhelpful in identifying a plausible causal structure that might link drivers of local growth with economic performance: do the hypothesized drivers enhance or retard local economic growth? Is the hypothesized relationship between the drivers and local economic performance linear, loglinear, or something else? Can drivers of local economic growth, such as 'social capital', plausibly be considered as causes of local economic growth, or do regions that perform well tend to attract 'social capital', or are they mutually determined?

To make any progress it is necessary to impose some additional structure on to the empirical model. This entails deciding not only what variables should be included in an empirical model, but also how these are assumed to be related to each other. Typically, this additional structure imposes an implicit causal structure expressing the nature and degree of dependence of one variable on other variables in the model system. In this instant, there exists an extensive and well developed set of growth econometrics that draws on recent development in the 'new' geographical economics (Durlauf et al. 2004). We have reservations about interpreting standard growth regressions in light of conventional growth theories. Nonetheless these relatively simple models do allow us to make causal inferences about the impact of local economic structure on differences in local economic performance using limited cross-sectional and geographically aggregated data. This allows us to compare our empirical results with an established body of theory and evidence.

Conceptually, growth regressions attempt to account for variations in economic performance across a set of geographical units. Essentially, these empirical models decompose the variability in growth rates across regions into three components: (1) adjustment dynamics with respect to an equilibrium growth path; (2) a set of (assumed exogenous) conditioning variables that determine the equilibrium growth rate for each spatial unit; and (3) a set of random shocks. Growth regressions postulate a simple 'gap convergence' relating local economic activity at the end of a given period (y_{it}), to the level of economic activity at the start of that period (y_{it-1}). Accordingly, the trend in local growth reflects, at least potentially, both a mean reversion towards a common steady state configuration of growth rates and differences in the local economic conditions prevailing between local economies. Formally,

$$\ln(y_{it}) = \beta_0 + \beta_1 \ln(y_{it-T}) + \sum_j \beta_j X_{ij,t-T} + \varepsilon_{i,t,t-T}$$

where $\beta, j \overset{K}{\underset{k=1}{=}} 1, \ldots, K$ are unknown parameters, $X_{ij,t-T}$ is the jth covariate associated with a local economy i at time t–T, and $\varepsilon_{i,t,t-T}$ defines a set of random and serially uncorrelated shocks to a local economy. In the context of this model specification, β_1 defines the speed of convergence in economic activity. In the absence of locally specific steady state disparities between economies ($\beta_2, \ldots, \beta_k = 0$), if $0 < \beta_1 < 1$ then local levels of economic activity will converge to a common mean, whereas $\beta_1 > 1$ implies divergence of local growth rates, and $\beta_1 = 1$ implies stable economic activity differentials.

Theories of spatial competition are supposed to provide a rationale for local convergence towards an equilibrium growth path in terms of the adjustment of prices, outputs, profit and wages to movements in commodities, capital and labour between local economies. The remaining systematic component, $\sum_j \beta_j X_{ij,t-T}$ represent the set of locally specific effects, reflecting the set of capacities in each local economy. In the broader literature, the selection of a set of social, political and economic conditioning variables remains poorly theorized. In the Australian case study, our initial specification of the set of potential conditioning variables was determined by how we chose to translate between 'institutional' theories and empirical measures. Whether these are actually drivers of local economic growth is an empirical question. Importantly, this model specification implicitly assumes that causation runs from the drivers of local economic growth to local economic performance. In this model specification, alternative causal structures are ruled out a priori. For example, it is difficult to imagine the level of social capital in 1980 being determined by local economic growth between 1980 and 2000. This type of causal structure is an attribute of this particular model, not empirical modelling *per se*.

Finally, the set of random and serially uncorrelated shocks to a local economy are supposed to constitute the non-systematic factors that account for variations in growth rates across local economies. Conventionally, this is denoted as the 'error' or 'residual' terms. In econometrics, it is common to interpret this residual as being derived from assumptions that we are willing to entertain regarding the types of error, approximation and omission that we made when constructing our empirical model: what variables should be included in or omitted from the model and how those variables are related to each other. In other words, the 'error' or 'residual' term is supposed to encompass the (in)famous *ceteris paribus* clause of economic theory. In my experience, the assumptions that we make about both the 'systematic' components and 'residual' components of our model are rarely met, and we need to consider how to proceed when things go wrong. Testing theories only makes sense if we have a correctly specified model. Accordingly, in practice we are rarely in the business of testing our theories. Rather, we are attempting to test our theories in conjunction with all the assumptions that we made in deriving our model specification.

In our Australian modelling exercise, the initial model specification, based upon a growth regression and a suite of theory-dependent conditioning variables, exhibited a systematic pattern of residuals. At this juncture, tacit knowledge becomes paramount. Once a problem has been identified, what should be done to correct the problem? Further inspection suggested that this specification error is most likely to result from a small number of outlying values. Should these outliers be removed, modelled, or should we consider the possibility that the data were drawn from a non-normally distributed population? Deciding on appropriate model selection strategies is fraught with difficulties

(McAleer 1994). We would like our final model specification to reflect the properties of the data rather than the strategy that we employed in moving from an initial to a final model. Hendry's (2000) methodology is helpful here, entailing a general-to-specific modelling selection strategy in which an initial general model specification is simplified to a final model that is 'congruent' with the theoretical and empirical information that is available.

Using this research strategy, we were able to empirically evaluate the suite of theories that we had operationalized using a suite of variables contained in our Australian data set. We found that, in the Australian context at least, our learning regions model out-performed competing 'institutionalist' theories of local economic growth. However, the set of drivers identified as significant are not those that are consistent with theory. Indeed, our modelling suggests that the processes driving local economic growth are much simpler than is typically hypothesized by the complexity and contingency of many 'institutional' theories. Specifically, we were able to identify two primary subsets of drivers of local economic growth: the local human resource base and the local enterprise culture. Given these caveats, the modelling exercise can be judged a success in so far as it provides us with evidence that it is possible to utilize quantitative methodologies to answer meaningful questions about the nature of the evolving economic landscape. While we may not be able to prove the absolute truth or falsity of competing theories of local economic growth, we were able to eliminate a set of false models against an alternative empirical model. We can tentatively accept this model for prediction and policy analysis until further evidence and/or different operationalizations are developed for empirical testing. In this context, there is value added by quantitative economic geography.

Quantification and methodology

Problematizing the challenges of bridging the gap between 'theory' and 'data' provides an opportunity to critically reflect more widely on issues of clarity, rigour and relevance in economic geography. Some critics of quantitative economic geography go as far as to deny the possibility of meaningful econometric modelling: we simply can't get at the mechanisms and tendencies operating in any local context; the world is just too complex and contingent to be captured by a relatively simple model; the only way to get at the essential causal mechanisms underpinning local economic growth is through qualitative methodologies.

Certainly, I would agree that empirical models are unlikely to capture the complexity of geographical reality. My view is that there is no guarantee that we will be able to discover the Truth, the whole Truth, and nothing but the Truth, regardless of our chosen methodology. All knowledge might be fallible, but not equally so. We could accept that anything goes, licensing all types of data-mining and data-snooping, but there are likely to be consequences for both theory interpretation and policy effectiveness. We may not be certain that we have the correct model, but we can eliminate poor models, at least as defined by our model design criteria. Here, I concur with David Hendry (1995: 10–11): 'the absence of a best, or dominant, strategy does not entail that "anything goes", since many strategies may be dominated even if none is optimal: try driving a car with your eyes closed.' From my perspective, the meaningful question to ask is: to what extent does a simplified empirical model permit us to account for the evolving economic landscape while simultaneously accounting for the errors, approximations and omission that we assume in our model specification?

On the one hand, the Australian case study provides evidence that the practice of quantitative economic geography offers more than its critics claim. In contrast with the stereotypical social constructions that are common in economic geography, quantitative economic geographers are not limited to searching for map patterns that need to be interrogated by qualitative geographers in search of causal explanations. Rather, we are using map patterns that have been generated using theoretically informed measurements on complex socio-economic processes to establish the empirical plausibility of competing theories, and assess the potential impact of policy interventions. These theories are based, either implicitly or explicitly, on causal assumptions that we make about processes driving change in the economic landscape. Whether, and how, causal relationships might be established using econometric modelling methodologies is currently subject to intense debate within the statistical sciences and beyond (Hoover 2001). Even in the non-experimental setting that characterizes most of economic geography, there does not appear to be any a priori reason why econometric models cannot be used to provide evidence for the operation of causal mechanisms.

Conversely, empirical modelling offers much less than the supposed 'ideal' of positivist/empiricist conceptions of inquiry demand. Even within its own terms of reference, there are practical limitations to implementing empirical modelling strategies that are imposed by the vague and ambiguous nature of available theories. There is no foolproof method for generating truth claims about the evolving economic landscape. Empirical results are contingent upon decisions that we make when selecting appropriate 'theory' and 'data' representations and in moving from an initial to a final model specification. Each stage in the modelling process is subject to uncertainty and potential error such that in the event that our model breaks down we can never be sure where the problem lies.

How might we do better? We might improve our practices by building better theories, collecting more comprehensive and appropriate data, and constructing more effective algorithms for bridging the theory–data gap. Recent developments in model specification and model selection have significantly enhanced such modelling strategies. However, improving the practice of quantitative economic geography means engaging with broader methodological debates in economic geography about social ontology and epistemology. Apparent incommensurability between methodologies creates barriers to understanding, reducing the scope for critical engagement with alternative research strategies. In this respect, recent debates in economics over the status and relevance of critical realism are proving to be particularly fruitful in transcending the supposed dualism between quantitative and qualitative methodologies in applied research (Downward and Mearman 2002). Following the dialogue initiated in Massey and Meegan (1985b), debates in economic geography assume the primacy of qualitative methodologies and ask the question: what is the role of quantitative methodologies? I prefer to look at the issue of constructing empirical knowledge of the social world from the other side of the coin, by assuming an economic modelling perspective and asking the question: how do 'intensive' case studies fit within an empirical modelling methodology?

Acknowledgements

I would like to thank Chris Fowler, Ron Johnston, Eric Sheppard, and Mike Taylor for their constructive comments on an earlier draft of this chapter. The usual disclaimers apply.

17 METHODOLOGIES, EPISTEMOLOGIES, AUDIENCES

Amy Glasmeier

This chapter examines a set of issues and challenges that face geographers wishing to engage in the policy analysis and the policy-making process. The discussion is not about research for research's sake, but about research targeted towards policy problems with contending parties and distributional consequences. In other words, where there are winners and losers. I make a further distinction between policy research that answers a specific problem in which the author chooses to follow a standard practice or recipe (impact analysis comes to mind) with neither alternative scenarios nor a range of effects specified versus policy research in which the analyst adds new information or incorporates theoretical propositions that go above and beyond or are in some way counter to the conventional view.

As part of this discussion I assert that activism, exemplified by Bennett Harrison's courageous life, requires the use of methodological practices that are seemingly in conflict with the qualitative turn in economic geography. Yet, instead of seeing qualitative and quantitative methods in conflict, I see the different approaches to research as complementary and use both depending on the problem at hand. Policy-makers usually accept qualitative research findings that are supportive of quantitative analysis and not independent of it. Discourse that pits qualitative research against quantitative research is self-defeating and deprives us of the power even simple numerical analysis allows for, especially in support of community groups and agencies involved in development.

This chapter has three parts. First, I discuss how politics and policy intersect and suggest how academics can use their skills as analysts to conduct research that runs counter to convention and which exposes ideological bias in policy debates. Second, I examine how conventional studies of firms and their strategic practices constitute important policy-relevant knowledge that can inform debates about how a changing economy affects geography and society. And third, I discuss a leading issue to which geographers can contribute, but for which there is little discussion of in the USA, namely, the consequences of growing inequality. Here I explore the role of intent in policy formulation and consider the role geography plays in the construction of social policies. I suggest that policy-making must be understood as a process that is inherently political and often 'irrational'. The formulation of, and therefore the consequences of, policy must be understood contextually specifically in terms of the circumstances surrounding the moment in time in which a policy is formulated. I conclude with comments about the role economic geographers can play in policy debates.

Points of departure

I finished my PhD at about the time the first *Politics and Method* (1985) volume was published. The American economy, like its counterparts in Western Europe, was suffering from a combination of the collapse of the Fordist economic model (Tickell and Peck 1992), the second oil crisis, the emergence of new competitor nations and the 'Voodoo economics' of the Reagan administration. The economy was undergoing a massive restructuring in terms of its geographical core, economic base and discursive power that was described as 'deindustrialization' (Bluestone and Harrison 1982; for a UK perspective see Massey and Meegan 1982; Sayer and Morgan 1985). Of course, the costs of such restructuring were disproportionately paid through job loss among working-class people who had once made a living wage in manufacturing and manufacturing related industries. The social consequences of restructuring were devastating as communities around the country experienced massive job loss and in some cases economic collapse.

Politics and Method centred on how to study the causes and consequences of industrial restructuring. The core debates were about what constituted evidence, whether underlying causes were better exposed and understood through extensive versus intensive investigations, how researchers of different political orientations chose to engage problems, the extent that research could be considered 'value free' and the types of policy prescription that were appropriate given the underlying causes of change. The emphasis was on interrogating problems, the consequences of methodological choices, and the role of ideology in shaping how researchers approached problems. There was little discussion of active engagement with groups that were either the objects of or subject to policy. Instead, at its heart, the volume explored differences in methodological approach in search of theoretical understanding – positivist claims versus structuralist interpretations of economic change. The chapters in the original volume were about how to conduct research by exploring the merits of different approaches to an understanding of the same problem.

In light of the debate of the mid-1980s, there is substantial evidence that economic geographers adopt many of the perspectives laid bare by the original discussion and in practice adopt a flexible attitude towards methodology responsive to the problem at hand. I count myself among this crowd. In that case I would call myself a methodological pluralist. Rather than being strictly a critic, I see myself as analyst that must make a choice each time I take on a project about how I will approach a problem and what the consequences are of the research strategy I undertake. As a researcher I am acutely aware of the consequences of challenging the model of 'normal science' embedded in policy analysis. I also firmly recognize the ideological bias of neoclassical economics and positivist science. At the same time, I do not set the rules of policy debate, but must work within them. My experience has been that the absence of conventional (this means statistical) evidence is a non-starter in most policy circles. Instead of conducting a frontal assault on a well-known approach to policy analysis, I see myself as using a variety of social science research methods to contribute to public policy debates.

The endless debate: economic geographers play in the policy-making process

An important contribution to the evolution of methodology in geography and social science more broadly is the recognition and acknowledgement of the situated nature of research and the positionality of the researcher (Fairclough 2003; Yapa 2002). This positionality goes beyond the specification of the problem and the role of the researcher in the act of conducting research and encompasses the role of the academic as the 'disinterested scholar'. Academic participation in policy debates is complex and different from when a researcher is gathering information in order to transform it into some type of scholarly report or contribution to scholarly debate. In the context of policy research, academics are treated as experts and are presumed to be tuned into larger social processes and able to offer some level of objectivity to situations. The distinction being made is between the scholar as the acquirer of information used to prove or in some way depict a hypothesis or research question and the scholar as information synthesizer and translator. This second role, may have greater burdens than, say, elite interviews as the researcher is not only bringing his/her skills to bear on information acquisition and problem understanding, but also in the process of information translation can and does determine what is then considered meaningful evidence.

In considering how to engage the policy-making process, it is important to appreciate what the policy-making process 'is', 'how it occurs', 'who does it' and 'why'. A couple of base-line questions are required when considering how we as academics, activists and geographers can engage the policy-making process. For example, a conversation in *Transactions* (Peck 1999; Banks and MacKian 2000; Peck 2000; Pollard et al. 2000) raised many important challenges for academics wishing to participate in the policy-making process. Yet, despite this, the debate seemed distant from the policy-making process itself. A point was made about the tendency for economic geographers to focus on the abstract (mental labour) leaving the more 'pedestrian' (manual labour) work of actually studying policy problems to others, including economists. A legitimate issue was raised about different types of policy and policy environment 'inside', meaning the university environment, versus 'outside' and who had the 'ear' of the prime minister, which I take to mean the wider world. The debate did not resolve anything in particular, but it did air issues that had heretofore lurked under the surface, such as the split between qualitative and quantitative approaches, the lack of credibility of policy research in the academy, and the decline in idealism in academic life. In that way it was very therapeutic. In the end, however, it didn't say enough about policy itself, how it is made, how research fits into it, and how we as academics can avoid the banal while still making important contributions to policy debates and the policy formulation process itself.

I see policy research as being problem-focused and empirical and for which there is presumed to be some legislative, administrative, or legal means to lessen or somehow alter the underlying circumstances or cause of the problem. Policy is complex and dynamic and consists of statements, processes, and measures for policy implementation. Furthermore, policy and policy-making is conditioned and shaped by the political, social and economic *environment*, as well as historical factors. Policy is formulated by actors affiliated with and representing organizations that operate through institutions to achieve policy implementation. Policy is made and implemented at multiple spatial scales: international and regional, national and sub-national (state, provincial or local). By implication, linkages and paths of influence between these levels are also significant for understanding policy.

Modes of engagement, methods of research

Activist academic

There are many ways to engage in policy debates and to conduct policy research. One is as an 'objective researcher' investigating problems in search of the 'right' answer. This is often accompanied by a pre-specified problem and a methodological approach defined by a 'client'. Another is as an activist working with interest groups to achieve some sought-after outcome. Often the first role bleeds into the second role depending on the issue at hand. For example, my first brush with policy, economics and the politics of knowledge arose in conjunction with research about a huge federal project being proposed for a remote part of the Nevada–Utah desert. Because of my background in impact analysis and environmental impact reporting, I was asked by an organization in New York to examine the Air Force's impact statement on the MX missile project, a mobile military weapons system planned for the desert of Nevada and Utah. I had previous work experience in the region. My early methodological point-of-view was shaped by data about projects that could be described by models and numbers.

There is a presumption of the neutrality of numbers. The MX project was no exception. The federal government had spent millions of dollars formulating a technical assessment of the effect of the programme on remote rural counties of two western states. The findings of the research were 'limited short-term effect and little or no long-term effect' on the region of the project. According to the Air Force consultants, their analysis was correct and little controversy existed regarding the initial findings of the federal evaluation. My research simply reanalyzed the numbers from the Air Force's technical reports using different initial conditions and assumptions about the process of development accompanying the project. For example, it was possible to find out how much cement was manufactured in the USA and where it was produced. This was easily matched with the project demand, which in turn was linked to the spatial distribution of cement production. Putting the information together, I showed that the project construction phase would consume all cement made west of the Mississippi for three years. My analysis indicated that the Air Force consultants had grossly underestimated the impact of the project and that it would forever change the region. Nonetheless, from evidence presented in public meetings it was clear the Air Force would shape the project's findings and use its authority, substantiated by ideology, to get the project built.

As a researcher I had finished my contract and could have walked away. However, my conscience prompted me to help local citizens contest the Air Force report and its analytical argument about the project's efficacy for the next two years. This effort focused on breaking down the information otherwise cloaked in technical jargon so that residents could understand the information used by the Air Force. Their problem was their lack of access to information and little prior experience with the National Environmental Policy Act, otherwise known as NEPA. Impact reporting as a tool is not in and of itself complex, it is largely recipe-driven. The key was to develop the ability of citizen groups to understand how the recipe worked and how the proportions of the ingredients were determined. For example, the Air Force impact analysis assumed the majority of workers would come without families. We found evidence to suggest otherwise, particularly if a time dimension was introduced (families travel with primary workers if the job is presumed to last more than a year). These alternative 'facts' were then incorporated into scenarios that spelled out what would happen in terms of demand for housing and other residential activities if the Air Force estimates were low. These results became important elements in the evidence phase of the public debate and the citizens' efforts led to an abandonment of the project.

This example demonstrates that 'truth' is a contested concept. Truth can only be known when enough of the variables or facts needed to know the truth are known. Missing facts affect the level of truth that can be shared with others. In instances when the level of uncertainty and stakes are high, truth will be hard to find and will more often than not be buried in politics. In such cases alternative points-of-view, based on evidence, are critical elements of analysis and can matter. Knowing data and knowing how to use them can be helpful to others. In this case a known method was used, a set of alternative scenarios implemented and counter findings inserted into the policy debate.

Academic practice applied to a policy problem: the study of firms

My second example highlights how our knowledge has substantial value in informing public policy debates about the consequences of economic change and processes of globalization. While the results of economic geographic analysis may not be able to affect the outcome, the findings from such research can be helpful in informing policy debates.

In the early 1990s, the North American Free Trade Agreement was being negotiated in Washington, DC. I was examining the early effects of globalization in rural areas. With the goal of informing the hearings and the Congressional vote to create the North American Free Trade Area (NAFTA), and to understand the extent to which local facilities and workers were aware of corporate strategy, we undertook a study of globalization. The research was conducted in three phases: first, using federal statistics, we mapped the location of industry employment and plants; second, we completed sectoral studies with attention to strategy and structure of the industries; third, we undertook corporate interviews with facility managers around the country in sectors likely to be affected by globalization.

Two levels of information were collected. The corporate strategic studies were based on secondary information collected from newspapers, trade journals, and sector studies compiled by academics and research organizations, including trade groups. This information was compiled into strategic profiles of sectoral responses to international trade conditions. In effect we were developing profiles of the probable actions firms could face given growing international competition. The facilities information was based on structured interviews that placed visited facilities in the larger corporate context. The interviews explored corporate strategic plans for coping with globalization and the extent of local knowledge of corporate plans for specific facilities.

Contrary to federal information and research based on statistical trends and static input–output analysis, it was clear from these sectoral studies that the passage of NAFTA would yield high unemployment in manufacturing industries faced with competition from lower-cost production facilities located in Mexico and other parts of Central America. Using international trade theory, economists argued that NAFTA would result in gains for all parties to the agreement. In contrast, our interview-based research suggested that there would be large job losses in the USA once freer trade occurred.

Of critical importance, our emphasis on structure highlighted an important issue. The impact of the agreement hinged on the actions of firms of different sizes. With over 100 interviews with firms of varying sizes, industry leaders and lobbyists, it was clear that US companies were going to move operations to Mexico. The more bifurcated the structure of the industry (split between large and small firms) the more likely the large firms would sell out the small firms by advocating for NAFTA. Many large firms were already planning operations abroad and fully anticipated benefiting from greater international trade. My economic geographer's training meant that I considered the industry structure: by knowing that firms of different sizes would have different objectives, it was clear that the big firms would dominate the political process – even if they were the numerical minority – and would sell out the small firms in the industry (Glasmeier and Conroy 1994; Glasmeier et al. 1995).

Policy research does not always have its intended effect. Having accurate information does not guarantee that it will be used either to make better decisions or to change the behaviour of affected parties. We were contacted by the International Union of Ladies Garment Workers and we showed how large brand producers and retailers were lobbying to pass the law (Glasmeier and Conroy 1994). Despite this evidence, the Union believed members of Congress would not pass the bill because of the myriad small producers located in their Congressional Districts. The day before the vote, my team got a request from the Union to help contact the largest private textile producer in the country, to get funds to pay for a last minute anti-NAFTA media campaign. But by this time it was too late and I learnt a powerful lesson – one cannot assume that, with information and knowledge, people or organizations will do the right thing (Schoenberger 2001).

You can have perfect information and all of the knowledge in the world but if the ideology of the actors blocks them from grasping the significance of the change under way, you will end up losing a struggle even if you are 'right'.

Context matters: understanding our role in making policy

Engaging policy debates rests on a fundamental understanding that policy is about politics. Research informs policy, but it does not take the place of politics and in some circumstances may be entirely ignored or used purposefully to fulfil a political goal sometimes opposite of its intent (Welfare Reform is a case in point). A desire to participate in policy discussions and debates must be coupled with a deep appreciation for the history of the issue in question and built on recognition of 'political intent' which is often paradoxically linked to the underlying policy problem. A case in point is the history of poverty policy in the USA and the role that geography played in its early formulation.

In 1990, with a Democratic president and Congress, there was at least some interest among politicians that disenfranchisement continued to be a very serious problem in the USA. There was coalescence between politics and economics: because the nation was once again 'growing' jobs, the public and politicians were feeling a bit kinder-hearted towards those who were more economically vulnerable. The growing inequality between income and opportunity served to reinforce the need for some type of political intervention. This confluence made it possible to once again bring up the issue of poverty. In the first months of the new Clinton administration, politicians and members of the executive branch did just that. Unfortunately, as quickly as the issue was raised, it became mired in the debate about welfare reform, which led to one of the most draconian policy about-faces in contemporary political history. Clinton was put on the defensive and forced to support welfare reform legislation that would strip decades-old policy theory of any belief that people were poor not because they chose this path, but because there were structures and institutional constraints like location shaping their opportunity set.

The history of poverty policy in the USA suggests that good ideas and the public's interest in social concerns have a half-life based on the number of years left before the next presidential election. Politicians care about the immediate and think very little about the future. Just as the issue of economic vulnerability was becoming relevant, it disappeared from the radar screen. In 2000, the election of George W. Bush shifted the tone further, as his belief was that the most important way to help the poor was to encourage (force) them to take responsibility for themselves by acknowledging that their own behaviour was the basis of their plight. For President Bush, the solution to poverty was work and self-restraint. Bush used effective rhetoric once again to demonize the poor. This notwithstanding, there is no research evidence to suggest that poor people are poor by choice. Indeed, in the USA most of the poor are working poor and in poverty often through no fault of their own (Glasmeier 2005).

In response to this re-emerging dilemma, I have been trying to reconstruct poverty policy history in the United States to understand what politicians were thinking about when they first ventured to construct policy programmes to deal with inequality and the role that geography played in the construction of both problem and policy theory. I wish to see what underpinned the policy programmes of the 1960s and to understand the intent of the politicians who created the Great Society programmes and subsequent policy modifications over the last forty years.

What I've found about purpose is that policy-makers may begin with good intentions but, as demonstrated in a reading of policy history, policies are nothing more than a series of compromises. They are often unacceptable, but they are compromises nonetheless. Thus, in looking at the history of poverty policy in the United States, one finds that race, class, gender, and age are ingrained in the discourse – almost like constituent elements in a chemical preparation. The original policies aimed at eradicating poverty in the United States were really also about getting blacks out of the rural South and into the urban North, sequestering them in ghettos, and then, if possible, sending them back to the South when they were no longer needed for jobs in urban industries. However, by the 1970s they were stuck geographically and then isolated further by the creation of racialized housing programmes and decrepit neighbourhoods that lacked safety, services, and functioning schools.

In the USA there is an abundance of academic treatments of the history of poverty policy (particularly, O'Connor 2002). There are fewer studies that capture the underlying goals that explicitly determine the spatial consequences of such policy. This requires interrogating the historical record and completing content analyses of political discussions in a variety of contexts. Reading the Congressional Record, policy reports from agencies, internal agency memoranda, and administrative law governing the issue forms a base of information in which the language of policy-makers is explicit with regard to their goals in enacting policy. Confirming intent is done using open-ended interviews with key informants who were policy analysts, agency personnel and politicians themselves. Newspaper articles can corroborate findings from the administrative record. The bottom line is that historical understanding requires archival research coupled with personal interviews with individuals engaged in the policy debates of specific times. It is through this type of detailed qualitative analysis that geographers can understand both the intent behind policies and the explicit role that geography plays in the interpretation of policy problems and programmatic approaches to their 'resolution'. Such analysis highlights the intended, and by implication allows for the interpretation of unintended, consequences of public policy.

Using discourse analysis allows us to be clear that the USA has never really tried to eradicate poverty (as opposed to improve the condition of vulnerable members of society) because a serious assault would require fundamental changes in socio-economic relationships. Based on the record there were, at best, four years (1965–69) in which the nation spent a serious amount of money (billions) to eradicate poverty to reduce inequality. Sadly, as quickly as we formulated the policies, Congress began to struggle with the White House, resulting in a stand-off among politicians and ultimately weakening public resolve to solve these critical social problems.

Understanding why policy both ends up as it does and fails to achieve its goals requires understanding of the political context in which policy is both constructed and implemented. It is important to realize that in the USA a problem can divide government and literally pit the president on one side of an issue for one set of reasons against the Congress on the other side of an issue for another set of reasons. In the end, we're left with a battle of wills and the use of legislative rules to win. In US history, a divided government has been the result, with surprising frequency. Any policy born of a divided government may fail not because of its inherent intent, but because of party politics, pure and simple.

For the past three years I have worked on a project accomplished in conjunction with the re-reading of policy history. Studies of policy history serve two purposes. They demonstrate how to study government and government practices and how to understand the kind of rationality that exists in the formulation of federal law. If one is going to 'do policy', one must understand antecedents to be able to decide whether a policy is rational or irrational. If policy is, in fact, not 'rational' in the sense that it is developed without emotion and fear, then to change it requires strategies that expose the 'irrationality' and therefore the false promise of its construction. It is far more powerful to demonstrate that in America we have never tried to eradicate poverty based on evidence than it is to say that the problem is too difficult to solve.

Conclusion

As economic geographers, our skills, intuitions, political sensibilities, and identification with others add up to the possibility of being public advocates and participants in public dialogues and discourses without choosing sides. Our power comes from the way in which we conceive problems and investigate their meaning. I have worked in government and I have defended people at the same time. The most fulfilling moments in my professional career are those in which I am able to both address the problems that I see as most crucial and help people who are affected by them.

We should think about how our reasoning skills and sensibilities might benefit others who can learn from our analytical skills to make claims about issues of concern to a broader public. In other words, geographers can be teachers who help others develop the skills to argue on their own behalf. We don't just have to be the experts who represent others. We can, in fact, be truly useful as practitioners of economic geography in a public policy context.

Section 4

Boundary Crossings: Mobilizing Economic Geographies

18 OUT OF AFRICA: HISTORY, NATURE, EMPIRE

Judith Carney

Every type of research represents a way of seeing. Behind each way of seeing is a story. Here is mine.

It may seem odd that in the affluent post-Second World War USA, a child could know hunger. That happened to me after a heart attack carried my father away at age 39, leaving my immigrant mother with five children under the age of six. We lived off a social security cheque whose wage earner had contributed only a few years of service. My mother had an eighth-grade education and few prospects. There were no relatives to assist us when economic free-fall began. One brother was chronically ill and his medical bills left us with even less. Near the end of one especially frigid January in Detroit, the utility company cut off our heat. We coped by sleeping huddled together under all our blankets in the living room. Great fun, thought each of us kids, for it fit our imagination of what camping must be like. But it was not so amusing when we did not have enough food to eat, which happened two or three times during our childhood. No act of the imagination can dull the pangs of hunger.

As the eldest, I desperately wanted to deliver the family from poverty. I excelled in school and studied hard to go to college. When I received a university scholarship, I was elated. But a month later, my mother was diagnosed with terminal cancer. She passed away three months afterwards, just six weeks before I started university. Just 18 and ready to abandon my education, I tried to get custody of my brothers. The judge refused, telling me my education was an opportunity not to be missed. My brothers were duly placed in an orphanage; I was devastated. A college education was within my reach, but I lost a family. I wandered in and out of majors as an undergraduate until the poverty and economic marginality that haunted my childhood drove a need to understand the world. After college, I travelled throughout Latin America, working along the way at whatever I could find. I met a lot of people in dire poverty. I witnessed their struggle to survive, to grow enough to eat despite the insecurity of not owning their land. These dignified people who so generously shared what little they had taught me what truly mattered. My interests increasingly focused on third world poverty and development.

I applied to Berkeley's graduate programme in Geography because of its cultural–environmental focus and regional specialization in Latin America. Coursework in cultural ecology led to my abiding interest in the subsistence strategies that the poor developed for farming the tropics without damaging the environment. In tropical environments across the globe such ways of life were facing unprecedented threat from development initiatives, promoted by autocratic governments that privileged cattle ranching, mining, and colonization projects. In rural Maranhão, Brazil, I saw the depth and versatility of local practices that sustained the environment and the cultures that thrived there. Farmers planted rice and other food crops for subsistence while harvesting many types of environmental resources for food and trade. Trees, such as babaçu, yielded valuable by-products. Women processed the nuts into vegetable oil, for cooking and sale, while the palm's fronds and wood served in home construction. Other wild plants contributed to subsistence, healing, and liturgical practices of Afro-Brazilian syncretic religions.

In Maranhão's Alto-Turiaçu region I encountered for the first time *quilombos*, the remote and isolated hamlets founded by runaway slaves.[1] Ambitious road-building programmes to the Amazon interior were bringing these settlements into the national orbit. While cultivating the same crops as neighbouring communities, *quilombos* safeguarded varieties of seeds that preserved a cumulative and multi-generational selection of desired qualities. The agricultural fields of *quilombos* were farmed jointly, a practice that dated to the difficult years of living as fugitives from slavery. Mutual labour for collective subsistence had strengthened the capacity of their ancestors to produce enough food to survive and remain free. For their descendants, collective work reinforced cultural identity. It helped them resist land seizures, as property was held in common as a community trust. Throughout the Alto-Turiaçu region cultural identity was rooted in the environment and cultural memory in the landscape.

What grabbed my attention in rural Maranhão during the 1970s was the struggle of the rural poor to stay on the land. State development initiatives, supported by aggressive road construction, were bringing cattle ranchers, mining interests, and land speculators to the region. They had big plans for the farmland of these poor people. As neither *quilombos* nor other small-scale farmers possessed formal title to the land their families had worked for generations, they had no legal standing to defend their claims. The military government and its economic allies viewed them as impediments to their goal of transforming the Amazon. The violence of centuries of slavery returned to the region in a new guise, as invading economic interests used guns and intimidation to seize their land. Many rural Maranhense saw their communities destroyed as systemic violence forced them to flee. They became part of the huge exodus of poor Brazilians who crowded into the Amazonian colonization projects that ultimately failed them.

My experience in turbulent rural Maranhão in 1977 gave me two important insights that have informed most of my later work. First, it gave me a way of seeing the world – the interplay of culture and environment, the relationship of place-based knowledge systems to subsistence strategies and environmental sustainability. However, cultural ecology as a research discipline said little about the role of economic change and policies that privileged one type of land use over another. Having witnessed the shooting of one impoverished man who was targeted for resisting efforts to encroach on his land, I could not be satisfied with academic discussions of the sophisticated farming systems that the mixed-race descendants of slaves had devised over the past centuries. The ranching and mining interests that used frontier justice to force these peasant farmers from their land were threatening an entire way of life. I could not remain silent about this central fact. Like many other geographers working in the tropics where military governments ruled, I turned to political economy and the literature on peasant studies for a more dynamic conceptualization (De Janvry 1981; Scott 1976; Wolf 1966). In this way, the accomplishments and struggles of rural peoples could be situated against the background of economic change and power relations. I was among the generation of geographers whose fieldwork in the South demanded a more robust theoretical framework, which would become known as political ecology (Blaikie and Brookfield 1987; Hecht and Cockburn 1989; Peet and Watts 1996; Robbins 2004a; Zimmerer and Bassett 2003).

There was a second insight from my fieldwork in Maranhão that I did not appreciate at the time. I had actually experienced a powerful introduction to the African legacy in the Americas. When I went to West Africa for my dissertation research, I began to think about African continuities, the cultural heritage of enslaved Africans on New World landscapes. The fieldwork experience on both sides of the Atlantic moved me to think about the Atlantic Basin as a historical-geographical unit. When I returned once more to rural Maranhão in 2000, it was in the context of realizing that many of the subsistence farming practices that I witnessed in the 1970s held profound connections to West Africa.

But all this was not yet clear when I left Berkeley in 1983 to begin my doctoral research in Gambia, one of the world's poorest countries. At the time, international agencies had newly pledged unprecedented funding to develop the country and lift rural farmers – 80 per cent of the population – out of poverty. The objective was to make the country self-reliant in food production by converting the Gambian wetlands to modern irrigated rice schemes. The production of two rice crops annually would bring development to rural Gambia and farmers would reap the benefits of selling their surplus to the urban seaboard population. But as rice was traditionally a woman's crop, I wanted to know how large-scale commodification of the subsistence staple would affect female labour, income opportunities, and land use. Suspecting that I might be witnessing an epochal agrarian transformation, I set out for fieldwork in rural Gambia.

Gender and agrarian transitions

I began my research with a survey of farming practices in the central region of the country, which was the focus of these new development projects. The wetlands surrounding the Gambia River had served as the principal zone of food production during the colonial period, while the rain-fed plateau became specialized in peanut cultivation for export. However, Gambian environments had undergone an incomplete process of agrarian transformation. Commodity production on the uplands had reduced the acreage of land devoted to rain-fed cereals (millet, sorghum, maize) but revenues from peanuts, as with African commodities in general, had declined in global markets. While colonial policies had encouraged a greater reliance on women's swamp rice production for food, the country still relied upon imports for half its rice consumption. Irrigated rice projects promised to make up the domestic shortfall, but would now extend cash cropping to the wetlands. A subsistence crop traditionally grown by women was being commoditized.

Gambian farming systems had been studied since the 1940s by a handful of excellent scholars. Their work provided an invaluable historical perspective on the role of rice farming in subsistence strategies from the colonial era (Dey 1981, 1983; Gamble 1949, 1955; Haswell 1963; Weil 1982). The development studies literature helped conceptualize the research by drawing attention to the role of the state, market, and class for understanding how power relations affect rural agrarian transitions. Another set of insights derived from the peasant studies research, which examined peasants as a class and asked whether specific peasantries were differentiating, the moral economy that enabled them to accept at times some labour exploitation, their resistance to labour intensification, and peasant-based political movements (Bernstein 1979; Conti 1979; Hyden 1980; Scott 1976). Yet another focus of scholarship informed the study. This was Amartya Sen's pioneering work that linked hunger and famine to policies that weakened the claims of the poor to subsistence entitlements (Sen 1981).[2]

Nonetheless, the theoretical training in political economy, peasant and development studies, and cultural ecology in geography had not said much about rural women, households, and gender relations. The debates on markets, states, and the peasantry as a class were shaped around discussion of commodities. Women's work and subsistence crops were placed in the background even though female labour fed the peasantry. As I left for West Africa, the role of rural women in underdeveloped countries was gaining attention among a handful of female social scientists (principally in development economics and anthropology) in the academy. Their research brought attention to women's role in rural production and reproduction; however, the discussion remained framed within the analytical category, class (Agarwal 1982; Beneria and Sen 1982; Deere 1976). There was little geographical scholarship at the time on the dependence of specific agrarian transformations on female labour intensification, perhaps in part because there were not yet many female geographers engaged in fieldwork in the third world.

While in the field, I was not aware of the household studies literature that was emerging from the critique of the Marxist preoccupation with class as an explanation of agrarian transformations. Feminist concern with reproduction and production led to renewed interest in the household as a category of analysis, especially the significance of inter- and intra-household relations for examining female labour demands, income opportunities, the social construction of gender, and patriarchy. Household studies revealed that many types of response to poverty shaped the peasantry as a class. The burdens placed upon women, and especially the intra-household variables that resulted in differential access to and control over resources, were at times quite significant for understanding how peasant households responded to agrarian change and the effect of economic transformation on patriarchal family structures (Berry 1984; Guyer 1981).

Fieldwork in the Jahaly-Pacharr irrigated rice project in central Gambia quickly revealed to me the significance of the household for understanding gender resource struggles. No other research perspective provided comparable insights on how a crop traditionally grown by women could come under male control when transformed by irrigation technologies. Women were denied benefits from their traditional rice land while new demands were being placed on their labour. The outcome certainly reflected prevailing power relations and the cynical actions of development agencies and government officials who viewed the project as a way to strengthen political support in the region. But something else was going on and its understanding demanded engagement with the household-based rural production system: namely, understanding the land-use categories that enabled the male household head to access female labour, centralize intra-household decision-making, and to assume control over the surplus produced by female labour.

With the collusion of government officials, certainly aware of the gender dimensions of the rural farming system, and the donors, who pleaded ignorance of the social issues, the project exploded into a storm of gender conflict.

A development scheme that proclaimed direct benefits to female rice growers had actually granted control of the plots to their husbands, while reducing women's access to fertile wetlands, diminishing their income opportunities, and augmenting their labour burden. Gender conflict became manifest as struggles over meaning in the terms that customarily had regulated female labour demands, land access, and crop rights within the household. With lip-service to the imperatives of 'family values', the project managers designated the irrigated plots 'family' rather than 'individual' fields. This cleared the way for male household heads to assert new claims to female family labour while simultaneously centralizing their control over surplus rice income (Carney 1993). In short, women were expected to labour year-round to produce a rice surplus their husbands would appropriate. When the first harvest was complete, the women who worked it received no benefits. There then followed a loud and unprecedented outcry by the women, who deemed the compensation for their labour to be unacceptable. They then refused further work in the project's rice fields without some remuneration in paddy. Some households conceded their demands; others did not. Like its predecessors, the irrigated rice project was failing because plot size depended upon the full participation of family labour, not just females.

I returned frequently to the project area until the mid-1990s, when policy reforms had delivered the final blow to Jahaly-Pacharr. International Monetary Fund (IMF) structural adjustment policies had removed the price support for domestically grown rice in favour of imports. The reforms and donor aid no longer paid lip-service to 'food security' as a development priority. Instead, 'comparative advantage' held sway while liberalization of the market would presumably solve potential problems of food scarcity. Cheap Asian rice consequently flooded the urban seaboard while the cost of fertilizer quadrupled. The market for Gambian-grown rice dissolved when price supports for the domestic product were removed, a process that effectively severed the linkage between rural producer and urban consumer. The project had never built a mechanized mill to process the rice, leaving this critical step and marketing to the entrepreneurs, who never materialized in the region.

Each visit left me distressed about women's deteriorating access to fertile lowland swamps, which were used to produce subsistence rice, some of which was marketed to pay for food supplements and the schooling of their children.[3] Under the banners of 'food security' and a 'woman's project', irrigated rice development had worsened their socio-economic position, an issue that grew increasingly of concern to women-in-development advocacy groups over that decade. I nonetheless saw a way to combine my academic skills with activism by encouraging implementation of a bottom-up rice development project for women. I devoted one research trip to studying the agro-ecological practices and micro-environments in which women traditionally grew rice. Here was an autochthonous knowledge system that did not depend upon imported chemical fertilizers, costly fuel oil, and spare parts. It built instead upon the expertise and accumulated *in situ* knowledge of rice culture passed down through generations of women. Placing land in the hands of a group of women, instead of individual growers, and providing them modest swamp accessibility, offered a sustainable form of development of direct benefit to female growers.

I approached an especially sympathetic donor representative at one of the sponsoring agencies of the Jahaly–Pacharr project and proposed an alternative strategy for a group of female growers who lost rice land in the scheme. Eventually a project did ensue around traditional tidal swamp rice production, but not where we had planned. Prevailing instead were government officials who had learned from decades of foreign aid that development monies provide the means to reward [male] political supporters. Instead of helping female rice growers, the small-scale water development project's *raison d'être* was to curry political favour with specific ethnic groups in a peripheral rice-growing region. I learned a valuable lesson in development: that donor funds often end up serving the political-economic interests of governments they support and that development aid has more to do with legitimating existing power relations and providing hefty salaries to consultants than with alleviating poverty.

Jahaly–Pacharr, like its predecessors, failed not for the presumed laziness of the peasantry and their inability to understand technology, as one European engineer I interviewed, asserted. The project did not succeed because rural people, men and women, were already burdened with work, and plot size was calibrated by family, not just female, labour availability. Interest in drawing men into the cultivation of what was deemed a 'woman's crop' compounded the error giving *de facto* plot control to male household heads, who instead of helping their wives in the fields, devoted their energies to peanuts and petty trade.

By the 1990s, there was growing academic interest in gender and agrarian development. I felt an integral part of the collective effort of female researchers from the North who endeavoured to interpret the effects of economic change on rural women of the South. In attempting to explain how male household heads in rural Gambia manipulated land use categories to deflect intensified labour burdens on to their wives, I never meant to suggest that patriarchy in sub-Saharan Africa was especially pernicious, the interpretation some gave my work. My goal had been to situate economic change within the context of foreign aid priorities and to show how client states used the funding to legitimize their political authority. This was part of the process that contributed to the impoverishment of rural African producers and the declining socio-economic position of both men and women. But I also knew that when the Jahaly–Pacharr project collapsed, local people would prevail, even if the experience of 'development' had left their communities scarred and their hopes diminished. I had reached the limits of what I could comfortably say about another culture and how far I wanted to extend my commentaries about gender relations in the country.

African rice and knowledge systems in the Americas

When I first went to Gambia, I wondered how an Asian crop had assumed such importance in subsistence. I read that the Portuguese had introduced rice to the country during their maritime voyages. But in carrying out the library research for my dissertation, I was astonished to learn that rice had been in West Africa for several millennia. This was not *Oryza sativa*, of Asian origin, but *O. glaberrima*, an indigenous species independently domesticated in the wetlands of Mali more than 3,000 years ago. From there, African rice diffused over a vast region, all the way to the Atlantic Coast and south from Gambia to Côte d'Ivoire. Innovative practices accompanied the journey of rice culture across West Africa to forested highlands and mangrove estuaries near the Atlantic Coast. The traditional rice practices I had observed among Gambian women represented then the legacy of a knowledge system rooted in antiquity with numerous varieties adapted to drought, salinity, and flooding (National Research Council 1996).

My study of indigenous rice production (which eventually included research in Senegal, Gambia, and Guinea-Bissau) led me to a literature review of European commentaries on African rice systems over the first centuries of the transatlantic slave trade. These accounts provided considerable detail on the micro-environments cultivated as well as the prominence of women in the cereal's cultivation, marketing, and milling. In my third year of teaching at UCLA, I read the work of two historians who attributed the origins of the rice plantation economy of South Carolina to enslaved West Africans already adept at growing the cereal (Littlefield 1981; Wood 1974). I realized that my own research on West African rice culture could add to this scholarship through a cross-cultural perspective focused on indigenous production zones and practices. The historical overview of European mariner accounts offered a methodology for identifying key parameters of indigenous African rice culture that may have diffused across the Atlantic with enslaved Africans from the region.

The result of my research many years later was *Black Rice: The African Origins of Rice Cultivation in the Americas* (2001). The topic demanded that I shift the scale of research from localities and regions in West Africa to the areas of the Americas where enslaved Africans pioneered the cereal's introduction for subsistence and its adaptation to diverse tropical environments. I was compelled to think at a different scale, the broader historical-geographical unit of the Black Atlantic (Gilroy 1993). While my initial focus examined the Carolina rice plantation economy, where historians had made a strong argument for African agency, the research on Atlantic rice history eventually returned me to Maranhão, Brazil, where the Portuguese modelled a rice plantation economy after that of Carolina in the mid-eighteenth century. Rice remained the dietary staple of their descendants whom I had met in the region in the 1970s.

I began to think of ways to follow the cereal's introduction under different colonial experiences. Fortunately, US historians had so thoroughly combed the archives that I was able to identify the principal primary materials necessary for interpreting the rice environments planted and comparing their techniques with those described in West Africa. But this work also demanded considerable familiarity with the vast literature on slavery. I found a way forward by focusing my attention on plantation food systems, the crops grown for subsistence and the period in which the presence of rice is documented. This research involved thinking about rice as a system of knowledge, the suite of practices that surround its cultivation as well as processing. I term this 'rice culture' in order to bring into relief both agricultural practices as well as post harvest processing, the way the grain is milled and cooked. The basic features of African rice culture then provided the methodology for a cross-cultural diachronic exploration of Atlantic rice history.

Cultural and political ecology again pointed the way for considering the flow of African knowledge systems in the context of forced migration, enslavement, subsistence choices, and cultural identity. While my intellectual horizons at Berkeley developed from the insights of Carl Sauer and his respect for indigenous Amerindian knowledge systems, I had been struck in my own fieldwork by the seeming reticence of geographers to similarly engage the contributions of enslaved Africans in the Americas (Carney and Voeks 2003; Parsons 1972; but see Sauer 1966; Voeks 1997; Watts 1987). Working in rural areas of the Black Atlantic made me want to address this historical silence. I wished to illuminate the role of slaves in pioneering adaptation to the New World tropics and in establishing plants of African origin in new physical and social environments.[4] The botanical legacy of enslaved Africans also included the recognition of plant genera of pan-tropical distribution known for their medicinal properties.

Since completing *Black Rice*, I have extended the analysis of colonial rice history to Portuguese Brazil and Dutch Suriname. The introduction of rice to Brazil dates more than a century earlier than the cereal's appearance in the colony of South Carolina, while efforts to produce rice in the Dutch colony of Suriname were underway in the same decade as the English colony (1680s). The grain is present as a subsistence crop on plantations during three colonial experiences, even though it was not cultivated in any of these European metropoles (Carney 2004, 2005). But the historical record on rice is principally engaged with the cereal's potential as a commodity.

While researching the onset of rice cultivation in northeastern South America (the Guianas and Brazil), I learned of the maroon oral history that holds an enslaved African woman introduced rice by hiding grains of the cereal in her hair as she disembarked a slave ship. Here was a view of rice that contrasted with scholarship on the Columbian Exchange, which focuses on the role of Amerindian and Asian crops on transforming global food systems and environments. Instead of crediting European men with plant diffusion, the maroon account profiles enslaved African women, thereby providing independent testimony and validation of the significance of females in rice culture across the Black Atlantic. But the Columbian Exchange literature says little about the diffusion of African plants to the Americas. In part this is because many plants of African origin are erroneously believed to have originated in Asia. But it is also due to the research neglect of subsistence crops, which would otherwise illuminate the contributions of slaves in establishing those of African origin, such as rice, millet, and sorghum.

It is often forgotten in studies of plantation economies that slaves grew the food needed for subsistence as well as the crops exported as commodities. Enslaved labour produced the subsistence staples that both Whites and Blacks consumed. The scholarly emphasis on plantation commodities has not yet brought sufficient attention to the plants slaves grew for food, which they established when granted the right to individual plots. In serving as sites for crop experimentation and seed exchanges, these plots served as the botanical gardens of the dispossessed. In this way, the enslaved perpetuated African crops in the Americas.

Conclusion

Thinking about subsistence over the past decades involved me in a deeper journey into poverty, hunger, and inequality. In writing this chapter, I am struck by how tenuous so much of this research seemed to me while I was in the field. Retrospection always suggests a grand vision in terms of what one sees and wishes to accomplish. I have had a great passion for fieldwork. It has been an enormous privilege to have the opportunity to live in different areas of the world and to know people from different cultures and perspectives. If I had my life to live over again, I would follow the same path because it led to discovery, self-awareness, and compassion. Perhaps the personal is not often evident when students read the articles and books academics write. Theories provide the way scholars talk to each other, but the vitality of the experience is often lost in the communication.

Fieldwork is inherently lonely, personal, and reflexive. I spent a lot of time alone and many times asked myself why I deliberately chose this experience. What was it that made me crave going outside my own culture and living among very poor people in Latin America and West Africa? Somehow stripping away the gossamer that surrounds life in technologically developed societies of the West and going where the distraction of machines and deadlines did not exist helped me see what was essential. I learned something of great value to me as a person and scholar. I learned to appreciate being alive, to ask large questions and to venture a search for answers, even when conclusions were only partial. I saw my own life and struggle out of poverty in a context where I was an extremely privileged person.

One thing that saddens me is that in my own lifetime I have witnessed a widening gulf between 'haves' and 'have-nots' between and within societies, including our own. I am much more aware of how commodity fetishism reproduces the *status quo*, appeals to our acquisitive instincts, and contributes to inequality and poverty, not to mention the political–economic rhetoric that promotes this injustice. I expected so much more of the world and hoped my own modest efforts would make a contribution to cross-cultural understanding and compassion for others.

Research is an open-ended process. As scholars we venture to different places to learn, but paradoxically we become students again. Learning about slavery and its underlying violence, and seeing the unresolved tensions that are denied substantive discussion to this day, has been a wrenching experience. It didn't happen to my forebears. In retrospect, I am grateful to the Africans and their descendants in the Black Atlantic for allowing me to experience their cultures. They opened my eyes and taught me to appreciate the cumulative wisdom of what bondage, resistance, and entrenched poverty has taught them. I have experienced loss of loved ones; perhaps that is what drew me to this research. But I was given the opportunity to acquire an education and honour the ones I lost. I want that same opportunity extended to every son and daughter of African ancestors wherever they live.

NOTES

1 Their plantation crops they had previously grown included rice, cotton, indigo, and coffee.
2 Michael Watts (1983), my adviser at Berkeley, made an important contribution here.
3 Rural Gambia is polygynous, with each wife largely responsible for the care of her children and food on the days she cooks.
4 Besides African rice, these include black-eyed peas, millet, sorghum, okra, hibiscus, and kola nut (Carney 2003).

19 'I OFFER YOU THIS, COMMODITY'

Vinay K. Gidwani

October 1993. I was at the Tribal Research and Training Centre (TRTC) at Gujarat Vidyapeeth, in the city of Ahmedabad, India, trying to track down the original forms from a landmark rural survey conducted under the Gandhian economist J. C. Kumarappa in 1928–29. The survey, of 998 households in the economically deprived sub-district of Matar in central Gujarat, resulted in a publication that is both a singular early contribution to agricultural economics in India and a stinging indictment of colonial economic policies (Kumarappa 1931). But its arguments rely on aggregated statistics; ethnographic detail is markedly absent – almost as if it might compromise the factual gravity of the claims. These missing details intrigued me – was it possible that the actual survey forms had survived the tribulations of time? I was dubious but my hopes were boosted by a colleague in Gujarat who recalled using the original forms in the early 1970s. I wasn't interested in a re-survey. My research objectives were both more modest and more ambitious.

I had arrived in Gujarat strongly influenced by a rich and eclectic peasant studies literature, particularly the theories of agrarian change associated with Karl Kautsky, V. I. Lenin, A. V. Chayanov, and L. N. Kritsman (c. 1926); as well as Eric Wolf (1966) and Teodor Shanin (1972). But these otherwise exemplary studies lacked an ethnographic and structured historical insight into intra- and inter-household interactions, and how they influence patterns of agrarian differentiation. By contrast, Tom Kessinger's (1974) monograph *Vilyatpur, 1848–1968* reconstructs agrarian change in a single place via the innovative use of archival materials and family genealogies. Understandably enough, its considerable micro-economic and micro-political insights come at the expense of theoretical exactness. Was it possible, I wondered, to combine analytic rigour with methodological virtuosity?

If I could trace descendants of a representative sub-sample of households from Kumarappa's original survey, and be privy to the household-level information gathered by his team in the 1928–29 survey, there was a distinct possibility, I thought, of pulling this off. So there I was on a scorching October morning, waiting for the gates of TRTC to open. It was a moment of anticipation and excitement, one of those singular markers by which the uncertain arc of fieldwork gets inscribed in memory.

Gujarat Vidyapeeth, the site of TRTC, was established in 1920 by Mohandas Karamchand (Mahatma) Gandhi as an alternative institution of higher education for Indian students. Gandhi imagined that the Vidyapeeth would pursue an anti-colonial pedagogical and political mission by emphasizing, among other things, a non-orthodox curriculum to be taught in the Gujarati vernacular rather than the colonizers' English. Here, serendipity steps in as the guiding hand of fieldwork. Gandhi, as it happens, had asked my grandfather, Assudomal Gidwani, to serve as the first *acharya* (vice-chancellor) of Gujarat Vidyapeeth, as I learned when I mentioned to a relative that I was to visit the Vidyapeeth. Once aware, I casually – even opportunistically – assumed that this would help me navigate the bureaucratic hurdles of a government institution. But my lineage meant absolutely nothing to the head clerk at TRTC. On the other hand, the fact that I was based in the USA and at the time spoke fledgling Gujarati and had the audacity to assume that I could waltz in expecting to gain ready access to *his* records meant everything. Over the course of the next few days he repeatedly rebuffed me: first I needed permission from the Director of TRTC; when I brought this, the key to the records room was missing; when I kept returning, he averred that the records room had been flooded this past monsoon thanks to a broken window. When I asked if I could nevertheless check he looked at me with contempt and said, 'I am the head clerk, don't you believe me?' His undisguised hostility, which at the time I could not fathom, mutely challenged me to do something about it. He bargained on the fact that in practice there was little I *could* do. He was right.

My deeply anticipated research project was defeated before it started. I had to scramble to recover from the setback, and for many years thereafter I would revisit my encounter with the head clerk with an amalgam of savage regret and resentment – and bafflement.

It took me a while to accept another narrative of why the head clerk at TRTC had responded so sharply to me despite what I took to be my unassuming manner. What I thought was a straightforward request may have struck him – in the absence of initial mediation and proper acknowledgment of his authority – as an affront. His hostility could be read as a subaltern critique of both my class privilege (my assumption that I had the *right* to access what I took to be public records at a government institution) and northern academic production (that I could use this information to advance *my* goals and career in the USA).

The rudiments of this critique are part of the gradient that connects researchers in the North to global field destinations in the South. The anthropologist Hugh Raffles describes this incidental conversation with Moacyr, a skilled and deeply knowledgeable Brazilian research assistant:

> At one point – we were talking about foreign researchers – Moacyr asked me how it was that foreigners could come here, get to know maybe one or two

places, perhaps a couple more if they stayed a long time, and then return to their universities, stand it front of roomfuls of students, and teach about somewhere called Amazonia. What he asked, with a mixture of bafflement, irony, and refusal, allows them to make such a claim: to pretend that they know *Amazonia*? (Raffles 2002: 11)

Moacyr displays wonder at how his vastly detailed and nuanced knowledge of Amazonia is under-valued in the global political economy of knowledge production. Although the tenor of his critique differs from that of the TRTC head clerk, their responses reveal shared subaltern disquiet about capitalist circuits of knowledge, where those who control means of production – credentialized northern researchers – profit most heavily.

I can now state my two principal claims. First, knowledge is produced as a commodity within a spatial division of labour that characteristically profits researchers in the metropole. Second, this production is indelibly marked by another connective geography: *a politics of translation that is at once a politics of transportation* (see Spivak 1986). To count as 'knowledge', information must be moved from the peripheries to a metropolitan location and be given recognizable form within prevailing disciplinary protocols and debates. These claims throw up a corollary argument, that research conducted from the North in the global South is necessarily *aporetic*: it must proceed, but to do so it must navigate blockages or non-passages which I gloss as the 'ethico-political'. The ethico-political marks zones of *liminality* where the prior certitudes of theories and methodologies are confronted by demands that cannot be anticipated or resolved a priori. Economic geographers encounter the liminal when formulating a research problem, during fieldwork, and when translating field research into written product.

By likening researchers in the North (including myself) to capitalist entrepreneurs in a global knowledge economy, I do not seek to doubt their sincerely-held intellectual and ethical convictions. That would be simplistic and disingenuous. Rather, I suggest that we researchers in the North are bearers of social relationships that are capitalist. To recall Marx via Laclau (1990: 9 original emphasis), 'capitalist relations of production consist of a relationship between *economic categories*, of which social actors only form part insofar as they are *Träger* (bearers) of them'. Thus, capitalists do not count as concrete persons, of flesh and blood, but as agents of capital. This capitalist, within a mode of production geared to the accumulation of value, has no choice but to relentlessly pursue profit – or perish. The advice given to young scholars – 'publish or perish' – seems to capture precisely this imperative.

Within the northern academy, a researcher's academic survival depends in large measure on the exchange value of her written product, which, like any commodity, depends for its 'realization' on its social use-value to the academic community. It must be 'consumed'. If the social use-value is judged slender or if there is over-production (too many similar papers in circulation), then the researcher's product is, in a very precise sense, *de-valued*. Use-value is always 'social', and in the academic context what counts as use-value is governed by the regulative ideals of the prevailing academic canon. As in the corporate world, so in academia: risk-taking can never be completely autonomous from the exigencies of capitalist 'value'. If anything, non-conventional work always runs the danger of being recuperated as avant-garde commodity: the latest novelty on the academic scene.

Where, then, should we look for the liminal moments of the 'ethico-political'? Nowhere, least of all in fieldwork manuals. Eluding anticipation and hence domestication by knowledge; they are only instantiated in the practice of research and achieve significance precisely in relation to the problematic of 'value'. They can only be recognized after the fact. Heuristically, I propose that these moments can be conceptualized as border crossings in two epistemological senses.

First, ethico-political moments designate unforeseen ruptures within academic circuits of 'value', which jeopardize its production or realization in circulation. Every researcher encounters liminal moments in the course of research which put his finely (or crudely!) wrought enterprise of information gathering and production into crisis. These junctures threaten the circuits of value that garner academic merit, or profit. Thus the term 'ethico-political' struggles to signify those moments when reason finds itself unable to furnish the normative criteria for a decision. But some decision *must* be taken – this is what makes it *aporetic*, it must be passed in face of a non-passage – and depending on its (unpredictable) effects it may become impossible to continue the research in its previously envisioned form.

Second, in the course of processing and re-presenting the field research for circulation and publication, the researcher may arrive at a critical realization – if he has not already – of the regulative ideals that operate him; indeed, of the political economy of uneven development that enables him to profit within the academy. Again, this is a potential moment of crisis, of liminality, where the future trajectory of 'value' is placed into question. The researcher may start to ask what it would mean to write with a primary commitment to extra-academic social use-values that diverge from – even actively reject – the circuits of exchange and academic reward.

Let me partially illustrate these claims. In 1994–95 I carried out approximately seventeen months of field research in central Gujarat as part of doctoral research. The bulk of my fieldwork was in Matar. In April 1995, campaigning for the Taluka elections was in full swing. Three candidates were in the fray – a politically experienced and reputedly crooked incumbent from the Congress-I party, belonging to the affluent but numerically small Rajput caste; a locally powerful *sarpanch* (head of village *panchayat*) from the landed Patel caste, supported by the Bharatiya Janata Party (BJP); and an ex-*shikshak* (teacher) of the Gujarat Vidyapeeth, belonging to the numerically preponderant Baraiya caste, who was waging an improbable campaign for Taluka *panchayat* chairman as an Independent. His main campaign pledge was to eliminate corruption and, in order to build his identity 'with the masses' (in contrast to his well-heeled opponents) he had cleverly chosen the 'bicycle' as his election symbol. '*Cycle ne vote aapo*' ('Vote for the Cycle') was his improbable refrain. Somehow, that year, his message, which villagers ordinarily might have dismissed as a disingenuous election promise, plucked at the lurking resentment of many villagers, particularly those from subordinate classes, who either felt disenfranchised by an unresponsive and corrupt Taluka bureaucracy (small and middle peasants from the so-called Kshatriya – Baraiya and Koli Patel – castes), betrayed by the unfulfilled pledges of development by the Congress-I (Kshatriyas, Muslims and Dalits), or threatened by the Hindu nationalist BJP (Muslims and Dalits). Local elections in India, like their national counterpart, are games of patronage that fracture constituencies into interest groups, which then have to be soldered into new hegemonic blocs.

One of the villages where I was doing fieldwork became bitterly divided. The coalition of Brahmins, Patels, Kshatriyas, Muslims, and Dalits that had emerged during the village (*gram panchayat*) elections in 1993 to prevent a coalition of Bharwads and miscellaneous castes (including dissenters from the majoritarian Kshatriya castes) from capturing the seat of *sarpanch* fell apart during the Taluka election campaign. The Brahmins and Patels defected from their coalition and formed a new alliance with the Bharwads to campaign for the BJP candidate. Meanwhile, the Kshatriyas, Muslims and Dalits put their weight behind the Independent candidate. During this, I was seated one afternoon at the *galla* (petty store) of a Patel acquaintance discussing the various candidates when a young Muslim labourer walked over to purchase *beedis*. My Patel acquaintance asked him whom he was planning to support in the Taluka election. The youth mumbled an evasive answer, saying he wasn't sure. The Patel shop-owner, a village-level BJP organizer, asked again. He was again rewarded with an inconclusive answer. At this point a young Brahmin landowner and self-styled BJP leader walked over to the *galla*. He confronted the clearly intimidated Muslim youth and pointedly asked him why the Muslims in the village were supporting the Independent candidate, and not the BJP one. Giving the youth no chance to respond, he went on to accuse Muslims at large of being unpatriotic and seditious, parading the familiar allegation that during cricket matches between India and Pakistan the Muslims in the village, as elsewhere, always cheered for Pakistan. Then, with unexpected vehemence, he proclaimed that all Muslims in Gujarat should have been 'put on the train to Godhra in 1947'. The frightened Muslim boy shuffled uneasily, avoided our gaze and meekly walked away.

Upon asking, I was told by the Brahmin BJP organizer that his reference to Godhra was to an incident during the Partition, when Muslims attempting to flee by train to the Pakistani border from troubled areas in Gujarat were slaughtered *en masse* at Godhra Junction in north Gujarat. This was a liminal moment for me. The Brahmin landowner was an influential presence in the village in which I was doing fieldwork, and opposing him in public could have imperilled my ability to continue with field research in a village where I had already 'invested' six-odd months. I had no idea what those consequences could be, but at *that* moment I could not stand by without taking a stand – either to let the moment pass by in silence or to confront my Brahmin inter- locutor. In that instant I experienced neither an epiphany – 'I knew then what I had to do!' – nor the steadfastness and prescience that is the gift of those with unshakeable convictions. I felt timid and alone, consumed by guilt, and acutely conscious of the looming decision. I decided to object to the Brahmin landowner's words, at which point he turned to me and told me that I was exactly the sort of weak and unpatriotic *Gandhivaadi* (follower of Mahatma Gandhi's ideas) who had done most damage to the country, by tolerating and, thereby stoking, Muslim sedition. My research was not severely impaired, although my interactions with the Patel shopkeeper and the Brahmin landowner subsided considerably thereafter.

On another occasion, in another village, the consequences of a decision were different. Early introductions – particularly when new and unfamiliar to an area – can have lasting consequences. Getting off on the wrong foot with an individual or a community can make it difficult to salvage relations down the road. Jan Breman, a Dutch anthropologist who has done pioneer- ing work on labour politics in Gujarat, notes that members of the domi- nant caste – ever keen to maintain control – exert a gravitational pull on outsiders, particularly unknown quantities like researchers: they approach them, invite them, query them, monopolize their time, try to plumb their purpose, attempt to fix their social co-ordinates, and if the outsider shows undue interest in village affairs, endeavour to influence him to their points of view and limit his mobility within the village. Patels are the dominant landed sub-caste (*jati*) in central Gujarat. I was hoping to pursue survey and ethnographic research in a multi-caste village, where Baraiyas are the most numerous but where political control lies with the Patels and their caste allies (Brahmins, Thakkar Vanias and Panchals).

My first introduction in the village was to one of the wealthiest Patel farmers, Kashibhai, the brother-in-law of a retired Gandhian politician, also a Patel, who lived in a nearby town and whom I had come to respect and like. I allowed myself to be introduced this way – it was a liminal moment of decision, when I saw fit to get a toehold in a village where I was inter- ested in working by the entry route just described. On that very first meeting

Kashibhai had invited Nandubhai Mehta, a Brahmin teacher in the village's middle school, to join us.[1] That same day I was also introduced to a person called Rajatbhai. The entire afternoon had been spent in the Patel *khadki*. When it had come time to leave I had been cornered by seven to eight Patel farmers who quizzed me at length – who I was, where I came from, why was I here in Ashapuri, what I planned to do, how long I intended to stay, and so on.

I was never able to gain the trust of the Baraiya community in Ashapuri. Poorer Baraiyas, in particular, either avoided me or displayed thinly concealed hostility when I tried to approach them. In the 14-odd months I spent visiting Ashapuri, on a fairly regular basis, members of poor Baraiya households only acquiesced to talk to me twice – under very revealing circumstances. Consider one of these. Walking to Ashapuri along the metalled road, I saw a youth – he must have been 15 or 16 – sitting by the roadside eating a simple lunch of *bajri rotlis*, *shaak*, and *dungri*. On impulse I asked whether I could sit down with him. He shrugged his shoulders, so I sat down. I asked him in whose field he was working. With a turn of his head, he indicated an adjacent rice field (which I later discovered belonged to Shamabhai Makwana, one of the few landed Baraiyas). The youth – call him Kanu – was shy, but seemed willing to entertain my questions. We talked about his family's economic situation (used to be good, but debts have piled up), relations between Baraiyas and Patels (seeing no one around Kanu informed me that Ashapuri's Patels always played at dividing the Baraiya community and taking over their lands), sources of credit (the afore-mentioned Rajatbhai is the largest money-lender in the village, almost half the Baraiya households were in debt to him, Rajatbhai borrows money at low interest from traders in Nadiad and circulates it in Ashapuri at high rates of interest), and so our conversations went until a motorcycle with two riders – Nandubhai Mehta in the driver's seat and one of the Patel farmers on the pillion – slowed down as they passed us. Nandubhai, looking at Kanu but speaking to me said, '*Shoon Vinaybhai, vaaton karo chho?*' ('What's up, Vinaybhai, having a chat?'). Kanu, who had been glancing around furtively throughout, clammed up. Soon thereafter he got up and said he had to go home. I asked whether we could meet again the next day. He nodded uncertainly. We fixed a time and place. He never showed up (and I never tried to chase him down).

One final illustration, an excerpt from fieldnotes dated 28 March 1994 (abstracted from Gidwani 2000), reveals my recurring dependence on Kashibhai (abbreviated as K.D.):

When I revisited the village of Ashapuri in late March of 1994 K.D. Patel once more extended a warm welcome. … Over the mandatory cup of tea I informed K.D. of my desire to visit the lower-caste quarters. He seemed bemused, … [b]ut as a good host he indulged my request. We walked over to the Rohit *vas* (quarters). The Rohits were surprised to see us. They ushered us into the courtyard that runs the length of their *vas*. A *charpoi* (cot) was quickly pulled out. We were asked to sit on it. The Rohits, including the elderly ones, sat down in a semi-circle around us, on the ground. I was deeply embarrassed, but didn't protest. ... The Rohits flooded K.D. with questions about the *mehman* (guest) he had brought. ... K.D. explained that I was a student from America, here to study the impact of canal irrigation. The children stared at me quizzically, the grown-ups skeptically. K.D. nudged me to ask my questions. Acutely conscious of the unequal basis of this exchange, I directed an apologetic stare at everyone and then inquired in hesitant Gujarati whether their *aarthik paristhiti* (economic condition) had improved with the advent of the canal. Most of the elderly Rohits nodded assent. But there were murmurs of dissent from some of the younger Rohits. They said wage increases had been offset by *monghwaari* (inflation). Work was unevenly available, with pronounced slack periods. K.D. retorted that work was always available to anyone who cared to work, although – he conceded – it was not always the preferred, physically less demanding, sort of work. He admonished the younger Rohits, stating that labourers these days were lazy, did not care to work, and expected to be paid for minimum effort. He vehemently denied there was any *berozgaari* (unemployment).

'I am always on the lookout for labourers (*majoor*), but often I have trouble finding any one willing to work. Labourers think I make them work too hard,' K.D. announced. ...

A tidily dressed man in the back, sporting a Gandhi cap, tilak, and white dhoti, bristled at this. ... 'How would you feel if you had to work in the field each day of the year for a living (*kamaani*)?' he asked. 'Wouldn't you feel tired, wouldn't you want to rest occasionally too, if your body was beaten in the fields day after day?'

'Don't labourers need rest? You big farmers only stand in the field and give orders ... it's easy for you to criticize the work habits of labourers.' The younger Rohit men and women murmured in support. The older Rohits seemed embarrassed. They tried to tell the man to be quiet, that this was a village matter. The man defiantly refused.

It transpired that the man was a schoolmaster from a distant village, ... visiting one of his relatives. We left ... after some more discussion. On the way out a few Rohit elders told me to come back again – but alone. I promised. K.D., who had retained his equanimity in face of the schoolmaster's accusations, simply said, 'This is what happens when outsiders who do not understand village matters interfere.'

That was the end of the matter for Kashibhai. I had stayed mostly silent during the exchange that occurred in the Rohit *vas*, despite agreeing with the sentiments of the visiting schoolmaster. I could have interjected while the exchange was occurring or raised the matter later with Kashibhai – who, despite his frequently retrogressive views on matters of caste (he once despaired that the caste system was breaking down and wondered, very earnestly, who would clear animal carcasses from village byways if Bhangis rejected their designated function within society), was otherwise an honest, extremely industrious man, who rarely dabbled in village politics. But I let the episode pass by unremarked. Why? I am not sure. Perhaps because right then I didn't feel entitled to an opinion on 'village matters' that I grasped only lightly (as Kashibhai had implied for the Rohit schoolmaster); or perhaps because I was afraid it would strain my relationship with Kashibhai, who had taken pains to teach me the details of *kharif* (monsoon) and *rabi* (winter) agriculture (an avocation for which he had a genuine passion and gift). In retrospect, the early associations with Nandubhai Mehta (the local schoolteacher who, I later came to know, was notorious for dishonesty and corruption) and Rajatbhai Patel (the moneylender who sometimes charged usurious rates of interest from borrowers) probably impaired my ability to gain the confidence of the Baraiyas of Ashapuri far more than my association with Kashibhai.

This brings me to the second sense of liminality: the ethico-political conundrums that mark the translation and re-presentation of lived labours of collaborators from the South in abstracted value-form. Two short clarifications are necessary. First, my use of 'value' draws on Diane Elson's (1979) reading of Marx. I take 'value' to imply those rationalizing practices whereby concrete labour is continuously enrolled and represented in abstracted form through the medium of money, as exchange-value. There is also an implicit argument here about space: namely, that the move from concrete to abstract labour entails a binding of distanciated events and occurrences into a common, interdependent capitalist space-economy, with money as the key intermediary. What is the academic correlate? It is precisely the 'truth' that a work of scholarship must not only circulate as exchange-value but also in order for its value to be actualized must be consumed as use-value (by journal referees, conference audiences, faculty colleagues, graduate students, etc. How else should we account for the importance placed, for instance, on counting citations?). Second, 'representation' implies for me not only the active and inevitable processes of interpretation, translation and writing entailed in the scholarly enterprise (re-presentation), but also the fact that representations that come to count as 'value' in the northern academy involve textual and territorial displacement from the subjects and subject matter re-presented. The researcher-producer is a representative, who is necessarily *dis*located from the subjects he claims to represent.

As mentioned at the outset, I embarked saturated in Marxist theories of the peasantry and pioneering Indian historical scholarship on place-based agrarian change, intending to carry out a genealogical investigation of social differentiation. Thwarted by the resolute head clerk at TRTC, my research morphed into an ethnographic and historical study of caste formation, agrarian relations and agro-ecological transformations. Reluctant to jettison my passion for and fixed investments in Marxist theories of agrarian change, I sought to combine the theoretical insights of Marxist geography with *marxissant* theories of cultural practice, and rational-choice models of land, labour and credit contracts based in the so-called New Institutional Economics (NIE).

What exactly did I expect to achieve? I had recently carefully read Lenin's masterpiece, *The Development of Capitalism in Russia* (1956 [1899]), investigating the transformation of a simple commodity economy into a capitalist one. The book's core thesis is established early: the 'basis of the commodity economy is the social division of labour' (Lenin 1956 [1899]: 11). Through an analysis of 1895–96 *Zemstvo* data he identified the four key mechanisms of economic differentiation within agrarian economies: (1) the ability of large farmers to exhibit greater *technical efficiency*; (2) to reap economies of scale in production; (3) to circulate cash surpluses as *usurious capital*, extracting interest rents from borrowers; and (4) to *extract surplus value* from wage-workers and utilize it to improve cultivation practices and further accumulate land. He never formally spells out the micro-economics of agrarian interactions, but new institutional economics does, choosing to understand contractual and social arrangements within agriculture as the outcomes of strategic interaction between rational agents under conditions of imperfect information and/or incomplete markets. Yet neither Leninist political economy nor NIE is much concerned with 'cultural practices' – the complex productions that give sense to the everyday lives of people.

I had also carefully read Bourdieu's *Outline of a Theory of Practice* (1977), coming away deeply impressed by his exposition of a culturally saturated theory of action. Bourdieu manages to stitch a Marxist–Weberian analysis of capital as a structuring field of power together with a devastating critique of Lévi-Strauss' structural anthropology. He offers an ingenious recuperation of rational choice theory that gathers inspiration from diverse sources: Maurice Merleau-Ponty's phenomenology of perception and Erving Goffman's 'microsociology' of public interaction. Bourdieu provides a rich tapestry of life in Kabylia – a rugged region of Algeria inhabited by Berbers – where social actors tacitly know how to act and interact in ways that serve to maximize their accumulation of 'symbolic' and 'material' capital. Actions are not analyzed as discrete units but as embodied practices that emerge from agents' 'generative dispositions', and are expressions of *habitus*. Bourdieu makes the provocative claim that these practices are disinterestedly rational: optimizing without agents necessarily being conscious of it. I thought, why not take Bourdieu's propositions and try to fill in the gaps in Marxist/Leninist political economy and NIE?

This was easier said than done, and I distinctly remember one incident – having returned to the USA for mid-research consultations – where an adviser, a noted development economist, looked at me as if I had gone mad when I told him, with great excitement, that I was trying to combine Bourdieu's practice theory of 'rational action' with NIE analyses of agrarian arrangements in order to come up with a fuller account of agrarian change. Bourdieu was foreign to his disciplinary canon and he plainly disapproved. It was a moment of passage when I realized that some disciplinary fences were going to be harder to cross than others. A decision loomed. I passed over into Bourdieu's territory, and although I remained on good terms with that particular adviser his interest in my project waned noticeably.

When I finally returned to the USA and, after a break, revisited my field-notes, I was struck by one particular entry from June 1995. It was a simple question: 'What am I doing here in Matar?'

That question contained, as I now realize, the germ of this chapter. Without consciously apprehending, I was already beginning to sense the limits of the stabilizing epistemologies to which I was then prone – deductivism and historicism, each of which mistakenly eases the task of research by positing historical change as a *logical* process. I was also beginning to experience acute disquiet around the conundrums of Northern research, where the demands of formalized, disciplinary knowledge generates an imperative to abstract and repair disjointed instances of the concrete into a coherent narrative that will be judged academically worthy.

I offer you this, commodity.

Acknowledgment

Many thanks to Joshua Barkan and Arjun Chowdhury for their questions and provocations, and to Eric Sheppard for editorial guidance.

NOTE

1 Names of persons and villages are altered.

20 'EL OTRO LADO' AND TRANSNATIONAL ETHNOGRAPHIES

Altha J. Cravey

My research seeks to understand the daily lives of workers who are swept up in globalization processes. I am currently focusing on Mexican workers who surreptitiously cross over to earn a living in the US, or as it is known in Mexico, *el otro lado* (the other side). Estimates reveal that one of every five working-age Mexicans has moved to the US for employment and the southeastern US has become a significant new destination. I have also documented the daily lives of Mexican workers in the *maquila* export sector in northern Mexico. While these *maquila* workers do not cross international borders, *maquila* factories rely on massive investments that *do* cross borders in a tireless search for profitable geographies. My research shows that these two groups of workers, those who cross the US–Mexico border and those who do not, are marked indelibly by gender and racialization processes that operate differently in various geographical contexts and that provide distinct choices to male and female Mexicans at the beginning of the twenty-first century. Mexican men – in far greater numbers than women – are drawn to expanding economic opportunities in the US. On the other hand, Mexican women have been the backbone of contemporary domestic industrial strategies. That is, from the beginning women have been predominant in northern border factories and women have contributed to a new style of industrialization that undercut and superseded previous industrial philosophies in Mexico (Cravey 1998, 2005b). How is it that expanding economic opportunities for Mexican men and women produce such disparate geographic patterns? How have politics and policies shaped these distinctly gendered and racialized outcomes?

A feminist approach focused on the interconnections between waged work and social reproductive work helps me to begin to unravel this paradox. Within this framework, I find that a transnational ethnographic method allows a deeper and more empathetic understanding of Mexican working-class lives and individual stories. Ethnography's insights derive partly through forming close emotional connections with a community and some of its members. In my experiences, the emotional aspect of ethnography is helpful for an analyst, yet at the same time, is fraught with difficult ethical choices. In reflecting on my own research practice, I know that I had many doubts as I started moving deeper and deeper into an ethnographic project a few years ago. The excitement of the project pulled me along, however, with such a force that I was caught up in the research before I fully admitted to myself what I was doing. Thus for me, a transnational ethnographic approach is somewhat new, yet I find ethnographic practice to be immensely worthwhile and stimulating. Furthermore, I find a transnational ethnographic approach deepens and complements case study techniques that I have used in the past.

My academic work is aligned with activist projects such as alternative globalization movements, anti-sweatshop activism, unions and worker justice organizing, and migrant rights advocacy. I take inspiration from these creative efforts to imagine alternative futures, illuminate social justice imperatives and improve ordinary people's working lives. My goal is to dedicate my research and pedagogy to explore, invent, imagine, and contribute to collective intellectual and activist efforts to create more equitable and just futures. Indeed, knowledge production is a central activity in social justice movements (Osterweil 2004).

Transnational ethnography

I use a feminist research framework that is attentive to the relationship between the researcher and the research 'object'. While there is no singular feminist method, there are a number of methodological implications that I grapple with in my research projects: situated knowledge, reflexivity, positionality, participant observation, and self-presentation. I elaborate these engagements and some of the challenges and limitations of ethnographic practice here, briefly discussing the relationship of my research activities and research agenda to activist and social justice projects.

In my current research, I use a case study approach that links participant observation, ethnographic methods, survey research and multiple secondary sources – what I call a transnational feminist ethnography. My focus is on the daily lives of Mexican working-class migrants in the US South. These migrants are predominantly male and mostly unauthorized, although about 40,000 enter the country each year on seasonal H2A visas. My project began with a community-based collaboration in 2000 and 2001 investigating pesticide exposure among rural agricultural workers who come from Mexico to harvest crops in the southeastern US. Community members, academics, and students interviewed hundreds of Mexican farm workers in Johnson and Sampson County, North Carolina, documenting their family and household arrangements as well as their understanding of the risks of pesticide exposure and knowledge of simple protective measures. In 2002, my own research focus shifted to urban workers and workplaces and to a time-intensive ethnographic approach. I immersed myself in the social spaces that Latinos construct in North Carolina, spending time in bars, nightclubs, pool halls, flea markets, festivals, and private apartments in the central part of the state. Through music and dancing, migrants – who are predominantly men – appropriate certain spaces as Mexican; find an outlet for sexuality, desire, and also for anger; and meanwhile distract themselves from the difficulties of living far from home.

In this chapter, I describe my fieldwork experiences in these 'Mexican' spaces and circuits as a way to draw out the significance and highlight some of the challenges of transnational feminist ethnography. I draw on various research projects in order to illustrate the politics involved in my evolving research praxis. My current fieldwork is transnational in the sense that some of fieldwork involves research in Mexico, and also because I closely engage with workers who cross back and forth from Mexico as a way of life. My own boundary crossings are thus caught up with the boundary crossings of Mexicans, and the translocal and transnational lives they create for themselves. I cross boundaries in social, and intellectual and other ways: such as figuring out how to interact socially with Mexicans in 'Mexican spaces'. The intellectual boundaries that I cross and hope to cross are the lines that separate academic work from activist work and advocacy. I describe these boundary crossings in more detail below.

As a participant observer in my current project, I keep myself open to possibilities that might open doors beyond my horizons and thus provide insight on issues concerning social reproduction, desire, and work. Because immigrant households are stretched across vast geographical distances, I have chosen to spend time in those places that seem to compensate, at least in part, for household relationships and activities that migration displaces and alters (Cravey 2003, 2005a). Presenting myself as a friendly partygoer, I have had the opportunity to participate in many social situations that might be closed off otherwise. In doing so, I have cultivated close relationships with a network of friends in central North Carolina and particularly close relationships with one extended family. In all of these relationships I have been quite open about my research interests and writing projects. While many conversations with numerous immigrants during this time have touched on pertinent topics, certain individuals were selected for in-depth interviews, either due to the specific situation that made their experiences compelling or, in three cases because they expressed an interest in having some input in 'the story'.

As a white, middle-class, middle-aged woman, I find the issue of clothing, grooming, and visual presentation to be a challenging aspect of this field project. Leslie Salzinger describes similar dilemmas in self-presentation and her 'new fashion self-consciousness' in her recent ethnographies of assembly-line work in northern Mexico (Salzinger 2003: 1). Experimenting with *mini-faldas* (mini-skirts) from the local thrift shops, tighter blouses, make-up, and contact lenses, I produced a 'look' that helps me to blend in at the free Friday night dance lessons in Chilango's restaurant and bar that precede free-for-all cumbia, bachata, salsa, and punta dancing. Adapting this 'look' for other social spaces meant studying the unfamiliar styles, preferences, and signals of others in order to spice up my usual bland repertoire and wardrobe. Self-presentation – down to the minutiae of clothing, mannerisms, and affiliations – can influence access to social spaces and specific individuals that otherwise exist in semi-invisible parallel worlds. The shortage of women in many of these locations is sometimes an asset, although at other times I attract more attention than I want. On most occasions, I find I can have more control and a better sense of my own safety by going out with friends rather than being alone.

Embodiment and lived experience are central to feminist theorizing. In my current research project, I struggle to link the body, embodiment, and lived experience to structural transformations such as the internationalization of labour markets in the southeastern US. While bridging such disparate phenomena is neither straightforward nor simple, the effort can bring together epistemology and experience in previously unimagined ways (Kobayashi 1994, 2005; Nelson and Seager 2005). Whereas adopting a new wardrobe and party-going persona might seem dishonest to some readers, for me it is an expression of a genuine long-term commitment to understand this community more deeply through an emotional connection as well as an intellectual one. Forging close relationships also allows me to observe the organization of under-recognized (and often unpaid) social reproductive work such as community and household upkeep, cooking, cleaning, caring for sick and elderly people, childrearing, and emotional work of maintaining intimate relationships (Kessler-Harris 1982; Mackintosh 1981, McDowell and Massey 1984). The gendered division of labour in these tasks provides a window on geographies of globalized production and the specific gendering and racialization of particular economic sectors.

Braceros and other border crossers

My current research project focuses on Mexican transnational migrants in the US South, a region which has been transformed in the last two decades by the rapid influx of foreign immigrants, the majority of whom are Latinos. A large proportion of the Latinos – about 75 per cent in North Carolina – are Mexican, predominantly from the working class. Of course, the movement of Mexicans to the US South is part of a long and cyclical history of Mexican migration to the US. The specific racial dynamics of the US South, and the racialized division of agricultural labour in particular, made the South the last region of the US to become attractive, and hospitable, to Mexican migrants. Regional economic expansion and aggressive promotion of the H2A guest worker programme (from the late 1980s) helped drive changes in spatial divisions of labour, producing a 'Latinization' of the region. The very rapid pace of change has created hardships for Latino immigrants and for certain communities that are changing quickly.

Thus, while Mexican migration to the US had surged in the First World War years and again in the Second World War years (via the bi-national Bracero Program, 1942–64), the demographic impact was highly uneven, and the historic plantation South was untouched by these labour flows until quite recently. As Edgar's story demonstrates below, however, these earlier movements from Mexico to the US had a profound influence on specific Mexican communities and on children and grandchildren who later decide to cross to the other side ('*el otro lado*'). Edgar Sotano Casteneda is a transnational *campesino* like thousnds of other Mexican men of the late twentieth and early twenty-first century. Agriculture had been a way of life for him until four years ago when, at age 71, he stopped going to the fields due to his declining health and began running a video arcade in his own home. His single transnational experience in 1952 opened the way for some of his children and grandchildren to become long-distance commuters in later years. In fact, the way I met Edgar is through forming friendships with his children and grandchildren, who were working in North Carolina, and talking to him and his wife several times on the phone before being invited to visit them in the state of Morelos. Through his family's experiences, I learned the extent to which border crossing becomes a way of life, entrenched in certain Mexican families in ways that sustain them and their communities.

When I talked to Edgar about his days working as a Bracero in the US, he showed me a few small black and white images that are carefully stored in a family photo album alongside snapshots of marriages, fiestas, and children. In the photos, from 1952, a 19-year-old Edgar stands casually alongside two other men, his posture and expression revealing a measure of pleasure and pride. He recalls that his mother begged him not to go to the US and cried when he left, yet Edgar said he was young and restless and just wanted to see what was there.

Edgar had heard about Bracero opportunities in casual conversations in his hometown and his interest was piqued by radio advertisements. To get a work contract, he travelled to Monterrey in northern Mexico on a commercial bus, and from there boarded a Bracero bus to a large contract location in Texas where employers where able to choose men for their farms. Edgar says he still remembers clearly how funny things were the first day. There was a 'big corral of men, as if they were goats', and the bosses picked out the men they wanted. His patron hired 300 men and drove them to a big metal barn on his farm that had hundreds of beds as well as a few stoves where one could cook and eat, and another section where tractors were stored. At the time, Mexican workers were paid by the pound for picking cotton and they put it in long bags that dragged along the ground. The boss also took them on a long trip into town so they could buy groceries. This part was even funnier, Edgar recalled. He felt really lost at the grocery store. 'It was so big and everything was marked in English. I had no idea what to buy.' After a few minutes, he decided to simply follow another man and buy whatever the other man bought. Later on, Edgar and four men from his home town decided to cook together, pooling their resources, and alternating responsibility for grocery shopping; one man knew how to make tortillas and another man fixed beans and food for their meals. He returned to Mexico in November of the same year, when it was becoming very cold in the panhandle of Texas where he was working. Edgar's border crossings and those of his children and grandchildren have become entwined with those of my own, and with my search for a deeper understanding of the daily lives of Mexican transnational commuters.

Crossing borders as an analyst

Transnational ethnography requires crossing cultural, intellectual, and national borders, each entailing its own distinct challenges for the researcher. For instance, Geraldine Pratt writes about those moments and modes of presentation that resist social scientific analysis because they defy understanding and cultural translation (Pratt 2004). Over time, a researcher may, or may not, gain enough discursive and cultural fluency to make sense of that which cannot be readily assimilated. Whatever the outcome, considerable time, effort, and resources can be invested in cross-cultural ethnographic endeavours with little guarantee of deepening understanding. High-quality ethnographic research cannot be accomplished without large investments of time.

I faced a different sort of barrier in pursuing the economic lives of mostly male Mexican workers in the US South. How could I – as a female – learn about their everyday lives and working experiences in North Carolina? Specifically, how could I enter the masculine world of the informal labour market just a few blocks from my own home? Would it be possible to pass time on the street corner with these clusters of men and uncover something meaningful about their lives? When I dropped by the convenience store (at the same intersection) and had brief conversations with these Mexican men, I noticed that the only women who seemed to fit in and belong were prostitutes passing through once in a while in search of clients. I never observed Mexican women at the street-corner labour market. How could I get to know some of the Mexican migrants at the labour market and hear their stories? Could I meet them in social spaces where they were relaxing and exchanging stories with each other? That is the strategy that I eventually undertook.

The day labour market is a place that is central to the lives of new immigrants in North Carolina. Workers cluster in knots in front of a tiny commercial strip alongside several apartment complexes. When trucks pull into the convenience store lot, workers quickly move towards the drivers to see who will be hiring workers. Even young Mexican men with dense social networks and full-time employment use the day labour market to meet employers and to pick up jobs on occasion. As a woman doing research alone, it did not seem like a good choice for field research, although I had a great interest in the place and considered various ways to approach it.[1] I could drop by the intersection, purchase a soda, have a conversation, yet I didn't really have a good excuse for passing my days there and, being female, would attract the wrong kind of attention.

Although I decided not to cross this particular boundary, I did create friendship networks with Mexican workers by frequenting a restaurant/bar at the same intersection. Friendly conversations helped me to learn more about the labour market, the kinds of jobs that migrants encountered there, and even led to occasional invitations to visit various worksites. On one such occasion, I gave some friends a ride to a temporary weekend job. My two friends were on loan from their regular employer, a landscaper, to a construction contractor for moving some gravel. Rather than explaining to the two men that he wanted the gravel distributed around the foundation of the new room addition and letting them accomplish the task, the contractor pushed them 'como burros' (like donkeys) all day long. He did this in a way that was at once helpful and dehumanizing. He got them busy filling two wheelbarrows while he positioned walkways for their trips back and forth to the foundation trench. At the moment one section of foundation was filled, the contractor would reposition the plywood walkways to lead to another section of the foundation trench.

In this way, these two Mexican bodies were pushed to their absolute physical limits, enabling the contractor to squeeze as much as possible from each brown body for the ten dollars per hour he had promised to pay. By using his own white body to position the plywood walkways and to smooth and rake the surface of the gravel, he forced the two labourers to stay with the tasks of scooping and lifting gravel to fill the wheelbarrows and pushing the overflowing carts to their destinations.[2] Some six hours later, at the end of this work, each man was covered in sweat and complaining of sore muscles. In this instance, deskilling operates at the scale of the human body, separating the division of labour between the body and the mind to an absurd degree. Mexican bodies are harnessed to Gringo brains so that each can pursue their 'specialty'.

In this kind of extreme manual labour, the body is repeatedly pushed to its physical limits. Through close relationships with a number of unauthorized Mexicans in central North Carolina, I have observed these kinds of stressful and excessive situations firsthand. I try to think about these pressures in terms of the body itself – how each body is produced elsewhere (Mexico) and how each worker must nourish, rest, and otherwise maintain health in order to continue working. It is also interesting to note that many immigrants consider their sojourn in the US as a time to push *themselves* to an extreme degree, as they focus on working and accumulating money. As Max notes, 'I'm going to make whatever I can while I'm here so I can expand my house [in Puebla]. The sooner I do this, the sooner I can return home.' Therefore, it is not at all uncommon to find that the immigrant body in North Carolina works more than one full-time job. In Paco's case, two full-time jobs in the hotel banquet service are co-ordinated because the scheduling of the shifts can be easily synchronized: 'When I finish the day shift at the Sheraton, I have about 20 minutes to get over to the Holiday Inn [for the evening shift].' For other Mexican workers, a full-time job may be supplemented with a part-time job such as an evening job in a restaurant or seasonal work in a factory. Jose says, 'I have to come home and shower before I go to the restaurant. The money from Jim Bob [the employer at his day job] does not stretch far enough.' Still others find additional work in an irregular pattern at the day labour exchange or in casual jobs working for themselves, doing such things as yard work, handyman tasks, painting and simple construction, as well as selling food items, phone cards, or second-hand cars.

Transnational ethnography is fraught with ethical dilemmas because of the power dynamics involved in the research relationships. In my fieldwork, the problem of self-presentation and positionality are salient and interconnected. Would unauthorized Mexican men accept me as a friendly-partygoer, or see me as a prostitute, a law enforcement officer, a social worker, or something else? How did my position as an academic affect the stories migrants would tell me? Would it alter my relationships with Mexican migrants in North Carolina to do small favours, such as translate official documents, visit insurance agencies, or provide a ride? As I made closer friendships, I began to worry more about the specific manifestations of the power differences involved. As a citizen with a high-status job, money, connections, and a sense of entitlement, I had much more power in each of these relationships. How could I maintain a sensitivity to these power differentials and avoid exploiting migrants' confidences? This ethical issue is one which never fully goes away but rather manifests itself in various forms at different times.

Another issue that I did not fully anticipate was the question of my own personal safety and what to do in dangerous situations. This issue arose in bars and music clubs where I observed many physical fights and sometimes unwittingly was to blame. Drinking and dancing and fighting in these Mexican spaces at first seemed to be inseparable activities, although I gained a deeper understanding of the role of fist fighting over time, especially through the insights of Jose Limón in his analysis of Mexican-American culture in Texas (Limón 1994). In the specific instance of working-class Latino bars in North Carolina, I found sporadic and short-lived altercations to be common. In a way, this experience of personal danger in my research sites sensitized me to the prevalent risks that unauthorized Mexican migrants face on a daily basis. Thus, my own concerns about personal safety and risk, while exaggerated to some degree, were a route to increased empathy for much more dangerous and consistent threats (of deportation, injury, isolation, and death) that dominate the lives of Mexican migrants in the US South. My border crossings into dangerous micro-environments give me a taste of something I may never fully understand.

Conclusion

I engage in feminist transnational ethnographic research with the intent of deeper understanding that may lead to transformation: of policy, national relationships, cross-cultural understanding, and improved community integration and empathy. All of these transformations are bound up with projects that endeavour to think about 'space' in a more deliberate way and acknowledge that social relationships involve and produce social differences *because* they are spatial (Massey 2005). Such theoretical elaboration implies a 'relational politics of the spatial', which can open new kinds of political opportunities (Massey 2005). Indeed, if ethics and responsibility can be understood as operating 'at a distance' in the same manner as they operate nearby, this geographic re-imagining could contribute to social justice goals.

I cross borders in order to document the lives of unauthorized working-class Mexicans crossing the US–Mexico border in search of improved livelihoods and expanded freedoms. I have documented Mexican resilience and creativity in maintaining transnational networks that sustain them, and that reproduce labour power at significantly lower cost. These observations have awakened me to the positive and potentially progressive force of new migrants' vast local and transnational networks. Mexican workers are keenly aware of the hazards of crossing borders and the finite limits of their bodies, yet put these same bodies at risk daily. For them, transnational commuting has become an important survival strategy. Some workers, of course, decide to settle in the US for the long term. Others come for a few years and use their earnings to improve their houses or launch a business venture in Mexico. Many struggle over these choices of where to live, when to commute, and especially over the issue of whether children should commute with their parents.

In my current and future projects, I want to forge stronger alliances with activist groups involved in alternative globalization movements, migrant rights advocacy and worker justice organizing, who are regularly engaged in activities that grapple with the challenges of 'responsibility at a distance'. While I do work with some groups, I want to cultivate more of these kinds of connections, and pursue social justice goals in a more consistent way by helping to forge broader linkages between academic and activist projects. Feminist transnational ethnography is one key tool I can use in this regard, as it is especially sensitive to the very transnational processes that create the juxtaposition, the 'throwntogetherness', of profound cultural differences (Massey 2005).

Ethnographic research practices have clear strengths and weaknesses for economic geographers who want to understand the social costs and cultural determinants of globalization. A key strength of an ethnographic approach is that it allows one to keep a close check on unfounded abstraction. However, without a commitment to theoretical elaboration, ethnography can lead to serious misunderstandings. Furthermore, such qualitative data can be enriched through careful connection to quantitative evidence and analysis (Lawson 1995). For economic geographers interested in global issues, a multi-sited and multi-method approach is capable of 'grounding globalization' (Burawoy et al. 2000: 341). Various sorts of border crossings help us, as academics, to understand the realities of working-class transnational lives, and the imperatives of pursuing social justice alongside those who are swept up in the volatile maelstrom of global capital.

NOTES

1 I also use the labour market as a self-guided field trip for students in my classes, who can take a free bus to the intersection, stop briefly in the convenience store, and return to campus. To my surprise, a student last term told me that his Latin Americanist Political Science professor had also required students to visit the site.

2 These carefully orchestrated trips with heavily loaded wheelbarrows are reminiscent of Taylorist Scientific Management techniques for moving pig iron in Bethlehem Steel Company. A workman named Schmidt was selected from a gang of pig-iron handlers and 'trained' to move four times as much pig iron as workers had done previously (Braverman 1974: 102–6).

21 RESEARCHING TRANSNATIONAL NETWORKS

Philip F. Kelly and Kris Olds

Economic geographers conventionally study particular places, *upon which*, *from which* and *within which* economic processes operate: the national economy, the regional cluster, the urban labour market, etc. Yet a distinctive feature of contemporary transnational processes is the extent to which they operate *between places*, creating intensifying functional linkages across space that compress time and (re)fold space (Sheppard 2002). Ulf Hannerz (2003: 206) makes a distinction between 'multi-local' and 'translocal' processes, the latter implying relationships and networks that integrate multiple sites into larger fields. Tracing these linkages makes transnational research distinct from comparative research.

Economic geographers have also turned increasingly to qualitative methods such as interviews, focus groups, and participant observation. But if the standards of rigour in qualitative research are set by the expectations of extended, place-based research in the anthropological tradition, this poses a dilemma for multi-sited and multi-scalar studies of transnational processes (see Gille and Ó Riain 2002; Gupta and Ferguson 1997; Hannerz 1998, 2003; Marcus 1995).

In this chapter we explore some of these challenges of conducting research when the processes of interest operate between places and across multiple scales, using examples from our own work. One involves the linking of labour market experiences through households that are stretched across global space by migration; the other involves the articulation of the developmental agendas of the Singaporean state with elite globalizing business schools from the West via various forms of partnerships and Singapore-based campuses. We identify four key themes with which we have grappled:

- the framing and scaling of the research process;
- methodological choices when analyzing the spatiality of transnational networks;
- the positionality of the researcher across transnational research sites; and
- the practice of conducting research at a distance.

Filipino migration and transnational labour markets (Philip Kelly)

Since 2002, I have studied the ties connecting Filipino immigrants in Toronto with their places of origin in the Philippines. These transnational linkages take various forms and have various consequences. Financial obligations to non-migrant family members in the Philippines may intensify labour market participation by immigrants in Toronto. Remittances to family members can affect labour market participation by recipients in the Philippines, as well as reshaping gender norms, attitudes to work, household economies and class structures. Information flows shape educational, training and career decisions in the Philippines, and gender, class and ethnic identities, forged in the Philippines, become invested with new meanings in North America and contribute to a high degree of labour market segmentation among racialized Filipino immigrants. These racialized meanings are, at least in part, rooted in the colonial history and cultural/political domination of the Philippines by the USA (Espiritu 2003). These processes indicate the emergence of a *transnational social space* tying Filipinos' labour market experiences in Toronto to continued economic, social and cultural linkages with the Philippines. Places become functionally integrated, such that embedded economic processes in one place cannot be understood without reference to the other.

Framing and scaling

At the outset, decisions must be made about the sites and scales of research: what are appropriate units of analysis and how will they be spatially defined? These require careful attention, as transnational processes entail the transcendence of sites and scales (Hannerz 1998). Most of the literature on labour markets and migration indicates two 'natural' scales of analysis – the transnational household and the local labour market. The household is seen as a functional unit, stretched across space for labour market and migration decisions, while the local labour market provides the context for employment. Yet such fixed scales can become problematic.

The household is at best a slippery concept, potentially including absent members, and possibly excluding residents who do not contribute to, or depend upon, the 'common rice bowl' (see Hart 1995). Transnational migration compounds this fuzziness, as contributing members of a household may do so from afar, while others may send financial help to multiple households without necessarily counting as members of any. Furthermore, the very notion of a household may vary across contexts – a problem specific to transnational research. In Toronto, the household can reasonably be equated with the 'house', but in the Philippines it is not uncommon for an extended family to occupy several houses clustered in a compound, to some extent sharing resources. A bounded definition equating the household with a 'home' ignores the possibility that there may be differing boundaries placed around 'the household' in different settings.

My initial conception of linked labour markets – Toronto on the one hand and a series of localities in the Philippines on the other as the pertinent scales of analysis – was also strained by empirical reality. Many informants also had relatives in the USA, or had themselves immigrated via the USA (or had similar connections with Europe, Asia or the Middle East). In the Philippines, migrants do not necessarily originate from, and connect with, singular 'sending' areas. Their families may be in various locations and many will have lived in Manila prior to migration rather than in their 'home' in the provinces. Framing the research question in terms of bi-local linkages between Canadian and Philippine labour markets foreclosed such complexities.

In short, the very process under examination disrupts the notion of a field 'site' as a territorialized entity. Defining the unit of analysis as the household and framing the transnational research process in bi-local terms inevitably neglected the broader web of relations that transnational migrants maintain.

Methodological choices

Qualitative interviews with individual actors formed a central component of my research design. Personal narratives are uniquely capable of accommodating the complex stories of migration and transnationalism (Lawson 2002), but are much better at evoking a phenomenon than determining its importance. There is a tendency in transnational research to research only those with transnational dimensions to their lives, but this risks ignoring perhaps the majority for whom transnationalism is not a daily reality (Portes 2001). Thus aggregated quantitative data is needed to establish the prevalence of the processes identified during interviews.

I discovered that statistical data served another purpose as well. I became involved with a new coalition of Filipino organizations in Toronto, the Community Alliance for Social Justice, which sought to campaign for the rights of domestic workers, for immigrant access to regulated trades and professions, and against racism and discrimination. It became apparent that my usefulness to this organization was based in large part on the statistical data that I could access and analyze. They had little use for qualitative narratives that might move a (white) academic audience, as such stories were essentially their own, but data showing high levels of education and professional credentials alongside low wages and a very segmented labour market was a greatly appreciated contribution to their campaign.

As a result, I spent more time than anticipated using and presenting such data, even though they provide little more than a broad context for my own research. Data generated within the scale of the nation state and its subsidiary administrative units does not fit easily with the transnational lives of migrants. By contrast, data on previous work overseas has important implications for settlement experiences, but was difficult to obtain except through personal interviews.

In order to extrapolate beyond personal interviews and say something broader about the processes being studied, I resorted to conducting my own surveys. I administered a survey of 300 households in a neighbour-hood in Manila, and another one in a rural village outside the city, to assess the local labour market effects of international migration and remittances. The Toronto alumni association of a Philippine university also assisted in conducting a survey of its membership to assess transnational linkages and employment experiences.

Surveys present their own challenges, however. The fixed language of the questionnaire inevitably frames the question at hand, but wording can carry different connotations in different contexts (above and beyond the compli-cations of translation). I encountered this issue when addressing the ways in which *class* identity was differentially conceived by Filipino immigrants in Toronto and their families left behind. In the Philippines, the consump-tion patterns that become possible with even 'working-class' employment in Canada, connote an upward mobility that is seldom possible without migrating. For those 'living the dream' in Canada, however, class may be defined less in terms of consumption and more in relation to their depro-fessionalized and deskilled status in the labour market, in which case their reading of their own class may be quite different from their relatives back home'. Transnational research deploys concepts that may be contextually specific. The fact that the group being studied in Toronto is defined by its origins doesn't necessarily imply that the concepts used in the research process will have the same connotations in both places. In order to address these collective understandings of abstract concepts such as class, I resorted to organizing focus groups. They permitted open discussion and reflection among a group of 6–8 people in a way that is seldom possible in an indi-vidual interview or survey.

Ultimately, then, the project evolved into a 'mixed methodology' (see Bailey et al. 2002). Qualitative narratives, statistical sources, survey data, and focus groups were used to address different dimensions of the process under examination. Interviews provide insights into the connections created by transnational lives; focus groups clarify the collective understandings of certain categories and processes; survey data allow the prevalence of the phenomena to be assessed; and statistical data serve to contextualize the process, albeit largely limited to the separate national contexts in which data are generated.

Positionality

Economic geographers, especially those engaged in qualitative work, are now well versed in the notion that researchers have a positionality that shapes their relationship with the research subject (see McDowell 1992; Pratt 2004). The interview becomes a social, cultural and political event based upon the ways in which respondents relate to this identity (see Dyck and McLaren 2004). Positionality is always context-dependent, and transnational research demands that the implications of presenting different 'selves' in different contexts be closely interrogated.

In Canada, my 'otherness' from my respondents was based upon my whiteness, my professorial status, and my work as a government-funded researcher. This had profound mediating effects upon the material I gathered, in ways that are difficult to untangle. One aspect that was clear to me on many occasions, however, involved the pedagogical, and even performative, presentation of Filipino-ness. As a non-Filipino I was assumed to have very little idea of Philippine political economy, history, or 'Filipino characteristics'. Thus when exploring identity transformations (especially related to class, gender and 'race') I was repeatedly informed about Filipino personality traits. These became an important part of my data (as they spoke to how informants chose to represent their identity), but will always be coloured by the social chemistry of the interviews and focus groups in which they were articulated.

In the Philippines, I inhabit the identity of a visitor from afar, the representative of the 'promised land', and a link to distant relatives (in some cases carrying gifts in either direction). While in Canada, I may be read as a representative of the bureaucracy that immigrants would treat with caution and perhaps suspicion, in the Philippines, my positionality is based far more on my whiteness and only secondarily on my status as a government-funded university professor. My position of privilege (and perhaps power) remains, but is differently constituted when away from 'home'.

Research at a distance

My 'mixed methodology' was both planned and the result of unexpected opportunities that arose in Toronto. Assistance from unanticipated sources, such as the university alumni association in Toronto, led me down unexpected paths. The participation of students and research assistants with particular personal networks allowed access to certain parts of the Filipino community, including extensive interviews with nurses, and with immigrants from the Visayan region of the Philippines. The interviews with nurses led to a more concerted focus on the transnational dimensions of labour market segmentation in the healthcare sector, matched by interviews with nursing school administrators in Manila. Another set of unanticipated opportunities arose from my involvement with Filipino activist and community groups in Toronto. These permitted insights into the community from regular social events, public lectures, rallies, film showings, informal conversations, e-mail correspondence, and community newspapers. All provided an ongoing source of information and understanding that was beyond the formal research design but invaluable as a source of ethnographic insight.

The research underwent a series of twists and turns as it unfolded, but all of these unexpected data sources and opportunities arose in Toronto and not the Philippines. This highlights how a transnational research design creates an unevenness of knowledge production when the researcher is based in one location and 'fieldtripping' to another. The richness of information available from 'being there' cannot be matched by focused data-gathering visits to other, distant sites, and serendipitous opportunities to take the research in new directions are largely a product of being *in situ*. Thus the (re-)direction of this project was inevitably shaped by concerns and opportunities generated in Toronto. The researcher is always 'in the field' (see Hyndman 2001), especially when dealing with issues of global interconnection, but it still matters greatly where we are located.

Global assemblage: constructing a global education hub in Singapore (Kris Olds)

Since 2000, select cities in Pacific Asia have formed or significantly deepened formal institutional linkages with a variety of elite Western (mainly American and European) universities, including three prominent business schools, via a variety of mechanisms, including free-standing campuses and research institutes, joint degree programmes, and consultancies. Between 1998 and 2006, twenty North American and European universities formed significant relationships with the Singaporean state. In the process a distinctive, continually evolving 'assemblage' is being created (Olds and Thrift 2005). For Rabinow (2003), assemblages are an experimental matrix of heterogeneous elements, techniques and concepts. This particular assemblage is being created by merging the rationalities, institutional structures, and technologies of government of foreign universities and the Singaporean state. It is producing experimental classroom designs, new training programmes for Government of Singapore officials, joint research centres, foreign campuses on Singaporean territory and new versions of academic freedom in a discursive cum-material reframing of Singapore as a 'global education hub' (Olds 2005).

Framing and scaling

My project links into, and is framed in relationship with, the ongoing debates about transnationalism in human geography and the social sciences (Dezalay and Garth 2002; Ong 1999). However, its framing and scaling was also spurred by urban wanderings in the late 1990s while teaching at the National University of Singapore. During regular journeys to my son's school, I passed a building site undergoing intensive construction. A large five-story building emerged in late 1999, and in early 2000 the INSEAD name was hoisted on to its side.

The emergence of INSEAD in Singapore, 10,700 km away from its Fontainebleau base, piqued my curiosity. I subsequently linked INSEAD's emergence to the unsettling of Singapore's higher education landscape during the post-1997 Asian economic crisis era. This crisis, and the rising power of China, was generating the latest in a series of crisis-related restructurings in Singapore. Thus it became clear that two scales were fundamental to the research design – the globalizing Western university and the city-state of Singapore. The university was viewed as a functional unit, stretched out across space for developmental reasons, while Singapore (as represented by the Singaporean state) was viewed as a co-author of the assemblage for its own developmental reasons.

Over the course of my first year (2001–02) at a new job at the University of Wisconsin, it became clear that these two scales – the city-state and the university – needed to be better contextualized in a multi-scalar manner. Singapore's positionality as a postcolonial global city in Pacific Asia, and regional basing point for transnational corporations (TNCs), fuelled its ambitions and developmental capacities. Business schools were the first foreign institutions of higher education to open the 'worm hole' (Sheppard 2002) facilitating the formation of global linkages. In turn business schools are one of four interdependent institutions (the others being business media, business gurus, and business consultancies) making up the so-called 'cultural circuit of capital' (Thrift 2005) that is anchored in the Global North. In this context of mobility and boundary crossings, relocation, and engagement with a variety of intellectual influences via the literature, two broad multi-scaled research foci emerged (see Figure 21.1).

Transnational research, then, frequently forces the researcher to work at multiple scales, and in somewhat different formats, in different parts of the world.

Methodological choices

My focus on the articulation of the agendas of the Singaporean state and the foreign universities led me to rely upon semi-structured personal

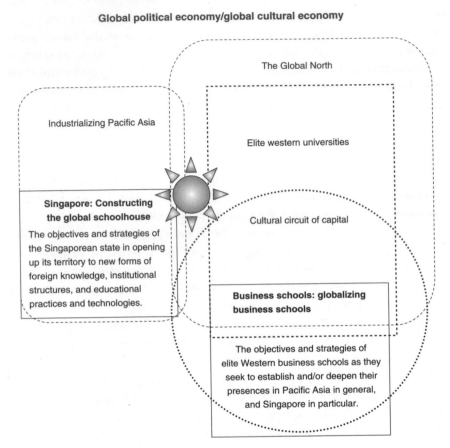

FIGURE 21.1   Research foci: global assemblage: Singapore, Western business schools, and the construction of a global education hub

interviews. This methodological choice relates to two factors: the nature of my theoretical influences – actor-network theory, governmentality approaches, and political economy approaches to the role of the state; and my appreciation for the role of narratives in the analytical and writing process (Czarniawska 1998). There is no pertinent standardized quantitative data in any of the territories in which I was to conduct research. Surveying the various knowledgeable actors was impossible because of the very small number of relevant institutions.[1]

The time pressures on senior university administrators and Government of Singapore officials, and the high-profile (and somewhat sensitive) nature of these initiatives, meant that interviews were the most effective single data acquisition mechanism. My respondents will rarely talk to me more than once, and face-to-face semi-structured interviews generate more trust, a sense of reciprocity, and information than any other method. In short, my focus on global elites, not global workers, favours more qualitative methodological choices.

Methodological choices implicitly privilege certain constructions of scale. Thus my dependence upon interviews led me to identify the importance of subsequently focusing on intra-institutional factors shaping business schools' globalization, including the nature of institutional hierarchy and their dependence on executive education as a revenue source.

Positionality

My status as a white middle-class Canadian academic, based in the USA but with work experience (1997-2000) in Singapore, facilitated my access to data. I am generally able to speak the same language, understand unspoken codes of convention, and engage in informal but important banter prior to and after formal interviews. I work in the same 'industry' as many of my interview subjects, enhancing my ability to approach them successfully, and engage in interviews. Unexpectedly, my University of Wisconsin-Madison affiliation was critically important in gaining access to key actors in all relevant spheres. University of Wisconsin-Madison is perceived as a prominent American public research university, and which all my research subjects had heard of and seemed to genuinely respect. Government of Singapore officials, who ignored me while at the National University of Singapore, now responded to my interview requests.

Brand-name university linkages proved insufficient, however, to overcome difficulties accessing busy university officials, or relatively cautious Government of Singapore officials. I attempted to redress these problems by obtaining modest financial support from the Social Science Research Council and the World Bank. I shamelessly used the names of these institutions, especially the World Bank, to generate attention and open doors to quantitative forms of data (e.g., internal programmatic data in relation to student make-up in foreign campuses and joint programmes). This enabled me to acquire valuable new forms of data during fieldwork in Singapore in May 2005. Interview subjects were particularly curious about the nature of the World Bank support I was receiving, and pleased that I might profile their institution in Bank-sanctioned documents. This arguably highlights that I am focusing on the apex (in terms of power) of the relational networks that collectively shape globally networked spaces (see Sheppard 2002, on power, inequality and global networks). Compared to Philip Kelly's experience with relatively less powerful agents, my interviewees make finer-grained assessments about my institutional affiliations (and my social and cultural capital).

Research at a distance

My research engages with 'the interlocking of multiple social–political sites and locations' that matter (Gupta and Ferguson 1997: 37). The emergence of a so-called 'global education hub' in Singapore is subject to interdependencies created within the broader political economy. Thus Singapore is the product of a constellation of social relations that stretch out across a range of scales (Massey 1992; Sassen 2001).

Given this conceptualization, and my own reliance on interviews, considerable effort is required to conduct research at a distance. It requires frequent travel to relevant sites, and funding is a challenge. Fortunately a wide variety of background documentation on the institutions I am researching is available in the University of Wisconsin-Madison library and on the internet. Thus I can spend less time in Asia than Philip Kelly, whose research subjects are not represented (or self-represented) in codified and publicly available texts. I can update the background data on a regular and relatively easy basis, despite having more sites to 'visit' than Philip Kelly.

Relocation to the USA facilitated and hindered the transnational research process. Despite stellar university resources, my commitment to the topic, my appropriate identity, and my affiliations with brand-name institutions, physical distance has made face-to-face access a continual problem. It would be much easier to conduct this research in Singapore, Chicago, Philadelphia or Fontainbleau. When frustrated, I remind myself that I effectively conducted four years of ethnographic research in Singapore between 1997 and 2001. I also reflect upon the relatively transparent nature of the institutions on which I am conducting research, and Hannerz's (2003: 211–12) caution that multi-sited research on these forms of global networks generally focuses on 'segments' of the lives of the ultra-mobile. In contrast to localized and deeply ethnographic studies, transnational researchers tend to create polymorphous forms of engagement (e.g., interviews, observation, analyses of media interviews, analyses of written texts) with our subjects.

Conclusions

Through brief accounts of two transnational research projects, we have highlighted some of the challenges that arise in all forms of social research, but also noted those issues that are unique to, or exacerbated by, a translocal research design that tries to track transnational processes.

Framing and scaling was a challenge. We began with specific, territorialized sites for empirical fieldwork, but it became apparent that the transnational ties we were exploring were part of more complex globalizing networks. We stopped short, however, of 'following the network', focusing our attention instead on particular nodes while remaining conscious of the expansive ties reaching outwards from them. This was a compromise between a 'place-based global ethnography', exploring multi-scalar forces at work in a single site (Burawoy et al. 2000), and a network based research design lightly touching as many linkages as possible (Hannerz 2003). We thus sought to recognize the ways in which global networks can transcend and rework fixed scales, while using field sites (local labour markets, campuses, etc.) as provisionally bounded scales. It seems inevitable that practical research will involve working with spatial categories in this way, while at the same time acknowledging their social constructedness and transcendence by empirical processes. This has the further advantage that the uneven power invested in different nodes in a network can be acknowledged (see Sheppard 2002 on 'positionality'). The household in the Philippines and the Philippine political economy more broadly are unequal participants in a network with their Canadian counterparts – indeed the entire migration process is predicated upon historical and contemporary uneven power relations. Likewise, although more subtly, business schools must acquiesce to the requirements of the Singaporean state, although ultimately Singapore needs them more than they need it.

The methodological corollary of studying selective sites in transnational networks was the use of interview techniques. Interviews allow the nuances of linkages and power relations within networks to be explored, and reveal dimensions that are simply not discernible in state-generated quantitative data. Quantitative data were, however, important for contextualizing or aggregating individual narratives and for the influence that 'scaled-up' data has in policy circles. In the study of Filipino migrants, therefore, a mixed methodology was used.

While it is inevitable that positionality will be performed distinctly in each interview, working across multiple transnational settings can establish quite different sets of researcher–respondent relationships. Kelly's 'otherness' from respondents was differently constructed in Canadian and Philippine settings. Olds' move from a Singaporean to an American university gave him a new, and more advantageous, positionality for accessing respondents in Singapore.

A more literal form of positionality relates to geographic distance from our research subjects. Transnational processes present the distinctive challenge that 'being there' for an extended place-based ethnography is almost impossible – not least because there are multiple 'theres'. Research at a distance inevitably becomes a part of the process. This is more easily achieved in some cases than in others. The formal aspects of university collaborations can be tracked through media and internet resources, but proximity is highly desirable when studying migrant lives and labour market decisions. When the researcher is connected more intimately with some research subjects than with others, there is inevitably unevenness in the research process, and the serendipities that arise in the course of research are one-sided.

NOTE

1 From Singapore, these were the Government of Singapore; National University of Singapore; Nanyang Technological University; Singapore Management University; and the local media (particularly the *Straits Times*). From the Global North, these were University of Chicago Graduate School of Business; US Government (American Embassy and Department of Education); Government of France (French Embassy); Graduate School of Business, University of Chicago; Wharton School, University of Pennsylvania; INSEAD; Association to Advance Collegiate Schools of Business; European Foundation for Management Development and business media (particularly the *Financial Times*).

22 REFLEXIVITY AND POSITIONALITY IN FEMINIST FIELDWORK REVISITED

Richa Nagar and Susan Geiger

> Some sort of reflexive identification of the academic writer with the 'Other' interpreted, analyzed or written about, is so important in reestablishing critical authority in the rubble of paradigms precisely because the most powerful and paralyzing aspect of the critique of representation has been its ethical implications for the very mode of communication – discursive, impersonal writing – so basic to academic work. (George Marcus 1992: 490)

Since the late 1980s, the practice of fieldwork has been under heavy scrutiny. The 'crisis of representation' – that is, doubts about the 'possibility of truthful portrayals of others' and 'the capacity of the subaltern to be heard' (Ortner 1996: 190) – has been particularly paralyzing in fieldwork-based research. Feminist social scientists in the *Western* academy, especially those focusing on *third world* subjects,[1] have responded to this crisis either by abandoning fieldwork, or by engaging in what Marcus calls 'a reflexive identification'. Here, we argue that neither approach can adequately respond to the challenges posed by the crisis of representation and that discussion about reflexivity, positionality, and identity – the central concepts in feminist interrogation of fieldwork – have reached an impasse. Simply stated, *reflexivity* involves a 'radical consciousness of self in facing the political dimensions of fieldwork and construction of knowledge' (Helen Callaway, in Hertz 1997: viii). In feminist conversations about fieldwork, reflexivity has often implied analyses of the ways that ethnographic knowledge is shaped by the shifting, contextual, and relational contours of the researcher's social *identity* and her social situatedness or *positionality* (in terms of gender, race, class, sexuality, and other axes of social difference), with respect to her subjects.

Here, we pose two key questions that lie at the heart of feminist research in third world contexts. First, how can feminists use fieldwork to produce knowledges across multiple divides (of power, geopolitical and institutional locations, axes of difference, etc.) without reinscribing the interests of the privileged? Second, how can the *production* of knowledges be tied explicitly to a material politics of social change favouring less privileged communities and places?

A widespread engagement with reflexive practices by feminist ethnographers has highlighted significant methodological and epistemological dilemmas endemic to fieldwork, as well as the challenges of identity politics as they affect academic, interpersonal, institutional and intellectual relationships. Such reflexivity, however, has mainly focused on the identities of individual researchers rather than on how such identities intersect with institutional, geopolitical, and material aspects of their positionality. This limited engagement has foreclosed opportunities for grappling with the two key questions posed above. Furthermore, existing approaches fail to distinguish systematically among ethical, ontological, and epistemological aspects of fieldwork dilemmas. Consequently, the *epistemological* dilemma of whether/how it is possible to represent 'accurately' gets conflated with *ethical* relationships and choices and with *ontological* questions of whether there is a pre-defined reality (about researcher–subject relationship) that can be known, represented, challenged, or altered through reflexivity.

Producing politically transformative knowledge across social divides necessitates that researchers rethink both the concepts and the ends of reflexivity, identity, and positionality. Below we describe how challenges in our own work led us to reflect on and articulate this impasse and elaborate two interrelated approaches – a *speaking with* model of research, and crossing boundaries with situated solidarities – for consideration by scholars who, like us, think that there is much to be lost by abandoning this face-to-face and necessarily problematic interactive research practice called 'fieldwork'. Richa's collaboration with a women's collective in India subsequently serves to suggest how these two approaches can be interwoven to extend reflexivity, and thereby to reconfigure the hegemonic models associated with fieldwork and knowledge production.

Articulating the impasse

This chapter began as an e-mail conversation in which we sought to articulate our struggles with field research for a dissertation (Nagar 1995) and a book (Geiger 1997b) in Tanzania. Our respective projects made us aware of contradictions inherent in a major recent trend in feminist social sciences. While feminists are committed to challenging pre-given social categories, an emphasis on 'positionality' requires reference to those very categories they seek to question. The embodiment of categorical identities in an individual is problematic because the social forces which name and reproduce them are continually in conflict with the necessary inauthenticity of the experience of that identity (Kawash 1996). In our cases, dismantling the homogeneous category *Asian* – constructed in a colonial context in East Africa – concerned Richa, while the categories and genders of Tanganyikans, supposedly responsible for anti-colonial nationalism in the 1950s, interested Susan. Ironically, validating our work to Western feminist academic audiences seemingly made it necessary to position ourselves using the very categories (middle-class, upper-caste, white, Indian, etc.) that we sought to problematize and disrupt.

There is a qualitative difference between an intellectual project which seeks to uncover complexity and a situation where an author feels that her work would be discredited if she is unable to establish herself as a legitimate researcher before her readers. For example, Richa received the following comment on an essay on marriage, migration, and religious ideologies among Asians in Tanzania from a feminist journal:

> Without knowing anything about the author, it is difficult to evaluate his/her use of the term *Hinduness* – a term that is probably insulting to the Hindus. ... Especially in the light of considerable work on interviewing, [the lack of a self-reflexive account] is simply not acceptable. ... Is the author male or female? The same goes for religion, caste and interviewing Hindus, Muslims and so forth ...

If Richa had merely revealed herself as a 'Hindu woman', this criticism would probably not have arisen. Yet, such a demand to uncover ourselves in these terms contradicts the purpose of problematizing the essentialist nature of social categories.

Gillian Rose (1997: 311) addresses this ontological problem in arguing that the 'search for positionality through transparent reflexivity' is bound to fail because it relies on the notion of a visible landscape of power. The imperative of transparent reflexivity assumes that messiness of the research process can be fully understood because it 'depends on certain notions of agency (as conscious) and power (as context), and assumes that both are knowable' (Rose 1997: 311). She (1997: 312) sees an inherent contradiction when 'a researcher situates both herself and her research subjects in the same landscape of power, which is the context of the research project in question' because 'the identity to be situated does not exist in isolation but only through mutually constitutive social relations, and ... the implications of this relational understanding of position ... makes the vision of a transparently knowable self and world impossible'.

Apart from this contradiction, we discovered at least three other problems. First, while many scholars expect ethnographic/life-historical research to explore the author's identity and positionality, no such expectation applies to quantitative methods. This unevenness implies that positivist research is immune to critiques of representation and, at worst, results in further marginalization of ethnographic research and personal narratives in producing knowledge. Second, to squeeze into a few pages multifaceted and changing relationships with our subjects in Tanzania or India entailed a translation that necessarily simplified complexity. For example, if in an essay on the politics of religious spaces Richa was expected to assess her complex positionality in relation to her subjects, could she also examine the religion- and caste-based relations among Hindus, Sikhs, or Ithna Asheris without reinforcing pre-conceived Western stereotypes of these categories? And if Susan devoted significant space to a discussion of her own positionality, rather than the situations and circumstances of the Tanzanian women whose political life histories she believed to be important, would she not be criticized for writing autobiography rather than a history of Tanzanian women and nationalism?

Finally, feminist field workers felt paralyzed when their efforts to be reflexive were read as 'rather puritanical, competitive assessment among scholars' (Marcus 1992: 490) or dismissed as 'tropes' that were sometimes seen as 'apologies', and at others as 'badges' which nevertheless failed to redistribute income, gain political rights for the powerless, create housing for the homeless, or improve wealth' (Patai 1991: 149; Patai in Wolf, 1997: 35;). Terms like appropriation, exploitation, and even surveillance are often attached to the very concept of 'Western' research among 'nosessment among scholars' (Marcus 1992: 490) or dismissed as 'tropes' that were sometimes seen as 'apologies', and at lly students, to conclude that they cannot step into 'other' worlds and societies for research purposes; or that it should not be done because it is inherently unethical (Geiger 1997b).

The next two sections explore ways to take positionality, identity and reflexivity out of misplaced struggles over legitimacy and transparent reflexivity and turn them into more meaningful conceptual tools that advance transformative politics of difference in relation to our own research agendas.

Speaking-with research 'subjects'

If the workings of power in fieldwork relationships cannot be fully understood, and the researcher's complex and shifting identities cannot accurately be captured, revealed, simplified, translated, or transposed across contexts, how can we approach these complex questions of power, privilege, and social change? One approach is a *speaking-with* model of engagement between researcher and researched – an approach that involves 'talking and listening carefully', and openness to influences of people from varied socio-cultural locations (DeVault 1999). These insights often remain abstract, however, requiring extension of reflexivity from an identity-based focus to a more material/institutional focus. Rather than privileging a reflexivity that emphasizes researcher's identity, we must discuss more explicitly the contextual economic, political and institutional processes and structures that shape the form and effects of fieldwork. Exploring the overlaps and disjunctures between these different reflexivities is essential for resolving the theory/praxis divide, engaging in feminist knowledge production across multiple borders, and moving beyond the impasse.

Feminists engaged in fieldwork are acutely aware of the dilemmas they face (Wolf 1997), but writing about these has largely been a post-fieldwork (even post-analysis) exercise in which the author critically reflects on the difficulties (including power differentials) pertaining to social relationships, reciprocity, and responsibility encountered in the 'field'. While this allows researchers to problematize the notions of bias/objectivity, it does not necessarily create spaces for critical reflection on the factors influencing the research process, its relevance to those to whom we are politically committed, the mutual benefits and enrichment exchanged, the *use* value of the work, etc. In other words, there has been little discussion of how to operationalize a 'speaking-with' approach to research that might help us work through negotiated and partial meanings in our intellectual/political productions. These questions must be addressed by thinking about reflexivity and positionality as processes.

Crossing borders with situated solidarities

The problem of voice (speaking for, to or with) intersects with the problem of place ('speaking from' and 'speaking of') (Arjun Appadurai 1988: 17)

Crossing borders is on the agenda. Foundations wish to fund projects connecting the global with the local, academics with practitioners. Editors are excited by cutting-edge scholarship that blurs disciplinary boundaries, emphasizes hybridity, and complicates the ideas of 'here' and 'there'. This is energizing but as Shohat (1996: 330) cautions: 'A celebration of syncretism and hybridity *per se*, if not articulated in conjunction with questions of hegemony and neo-colonial power relations, runs the risk of appearing to sanctify the *fait accompli* of colonial violence.' For feminists simultaneously located in institutions and communities of both North and South, it is critical to ask what borders we cross, in whose interests, and how our practices are interwoven with processes of imperialism and neo-colonialism.

Feminist fieldworkers have variously grappled with traversing borders, especially those separating the first and third worlds. Ong (1995: 367) finds it necessary to 'describe a political decentering ... in Western knowledge as it allows itself to be redefined by discourses from the geopolitical margins'. Such redefinition involves 'a deliberate cultivation of a mobile consciousness', engaged in a dialectical process of 'disowning places that come with overly determined claims and reowning them according to different (radical democratic) interests' and of 'critical agency shifting between transnational sites of power' (Ong 1995: 368).

Those of us who believe that the intellectual and political value of engaging in fieldwork across borders outweighs its problematic context (global capitalism, northern imperialism, structural inequalities), are responsible for developing critical analyses of our multidimensional struggles with such crossings. Jacka (1994: 663), self-described as a 'White, urban Australian woman', does this by describing the questions and criticisms received from Sinologists and Chinese men on one side, and feminists and postcolonialists on the other, as she carried out research on changing gender relations in rural China. Drawing on 'Western' theories and a decade of 'studying Chinese and things Chinese' Jacka (1994: 665) asserts:

> ... my focus on work and work relations derives as much from my interactions with Chinese women and from the Chinese Marxist emphasis on these issues as from western feminism. ... from a dialectical shifting between Chinese and western approaches and an attempt to synthesise the insights of each. This moving between cultures, gaining insights from both, is ... one of the most valuable aspects of the kind of research I am engaged in.

She argues, then, that there is no 'insurmountable gap' between her research and 'the concerns of rural Chinese women', not because all women share common oppression or a common set of needs/values, but because she recognizes 'the danger of essentialising the differences between East and West'. She insists that her research has been centrally shaped by the concerns and thinking of women with whom she has worked, and refuses to back away from what her research findings suggest:

> I *do* think that women in [rural] China are doubly exploited by the peasant family and by socialist patriarchy. ... And ..., while this, no doubt, does reflect my concerns as a westerner, I think it also reflects the concerns of Chinese women. In addition, I maintain that while my work may contribute to western discourses on the backwardness of the non-western, non-modern world, it also offers something more positive. ... I think western feminists have much to learn from the experiences of women in eastern societies, and, as Gayatri Spivak asserts, not only must White women do their 'homework' and learn about other women's experiences, their responses to other women's experiences will also allow White feminists to interrogate their own speaking positions. ... I recognize that, regardless of my individual motivations, in terms of world power relations, I work from within a dominating and colonizing discourse which imposes western, first world values on others, not the least by defining them as 'non-western' or 'third world'. I would hope, however, that we are not all completely trapped in the 'first world' or the 'third world' – that there is some possibility of developing counter discourses by sharing and working with others, by repositioning ourselves at least temporarily, both literally by doing fieldwork and metaphorically, and by 'smuggling ideas across the lines'. (Edward Said, in Jacka 1994: 667)

This kind of effective participation in border-crossings necessitates a processual approach to reflexivity and positionality, combined with an acute awareness of the place-based nature of our intellectual praxis. The goal must be to build *situated* solidarities, which seek to reconfigure our academic fields in relation to the 'fields' that our 'research subjects' inhabit. Similar to Dean's (1995: 69) 'reflective solidarities', situated solidarities aim to understand the larger interconnections produced by internationalization of economies and labour forces while challenging the colonialist prioritizing of the West. They are attentive to the ways in which our ability to evoke the global in relation to the local, to configure the specific nature of our alliances and commitments, and to participate in processes of social change are significantly shaped by our geographical and socio-institutional locations, and the particular combination of processes, events, and struggles underway in those locations. As Larner (1995: 177) suggests, it is inadequate for us to position ourselves only in theoretical and ideological 'place'; we must also recognize our 'geographical location, and by implication, the politics of that place'. Through her involvement in feminist discourses of difference among Maori and Pakeha women in Aotearoa/ New Zealand, Larner recognizes that there will be multiple situated knowledges rooted in different (and often mutually irreconcilable) epistemological positions in any particular context. Thus, the measure of contemporary feminist theory can neither be '"Who is making this theory?' nor 'What is the epistemological basis for the theory?' but rather 'What kinds of struggle does it make possible?'" (Larner 1995: 187).

Importantly, Larner maintains that despite the inevitable disjunctures, working with an understanding of positionality involves developing theoretical and political frameworks that integrate conflicts and contradictions:

> The goal is not unity based on common experience, or even experiences, but rather some sort of a workable compromise that will enable us to coalesce around specific issues. ... [I]f positionality is to be more than a recitation of one's personal characteristics, or a textual strategy, [we must address the dilemmas that] arise out of ... the politics of negotiating not just multiple, but discrepant audiences. Moreover, it may be that out of such engagements will come alternative theorisations, generated not out of abstract discussions about theoretical correctness, but rather out of the efforts of academics who are engaged with, and speak to, specific political struggles. (Larner 1995: 187–8)

This, we would add, requires an ongoing dialogue across geographical and disciplinary borders on how alternative theorizations can emerge from speaking to specific political struggles, and how inevitable disjunctures can become necessary pieces of reconfiguring the modes, purposes and meanings of knowledge production. The next section considers these questions by drawing on Richa's ongoing collaboration with the Sangtin collective in Uttar Pradesh, India. But a tragic irony surrounds this: as the text considers the possibilities and limits of a collective 'we' in Sangtin, the 'we' of

Susan and Richa is impoverished by Susan's untimely death in 2001, and replaced by Richa's 'I'. Yet, the collaboration with Sangtin was grounded in the terrain of reflexivity, positionality and identity traversed with Susan (Nagar 2006).

Reimagining reflexivity: notes on a journey-in-progress

In 2002, nine feminist collaborators came together to begin a public debate about the NGO-ization of grassroots politics and knowledge production in the Hindi-speaking NGO circles of North India. Seven of these authors – Reshma Ansari, Anupamlata, Vibha Bajpayee, Ramsheela, Shashibala, Shashi Vaish, and Surbala – earn their livelihoods as village-level NGO workers in the Sitapur District of Uttar Pradesh. Of the remaining two, Richa Singh works as a district-level NGO worker in Sitapur, and Richa Nagar teaches in Minnesota. The collaboration was undertaken as a way to imagine the future of Sangtin, an organization that the activists from Sitapur had co-founded with other local women to work for the self-empowerment of the socio-politically marginalized communities of rural Sitapur. Elsewhere, I reconstruct a chronological account of the evolution of my relationships with Sangtin and its members (Nagar 2002; Singh and Nagar 2006).

Here, I discuss the fragile nature of the 'we' in Sangtin and the possibilities created by dialogue and ongoing self-critique in a transnational collaboration. The journey began with our writing and publishing *Sangtin Yatra*, a book in Hindustani, which interbraids autobiographical narratives of the seven rural activists to highlight how caste, class, religion, and gender enmesh in the processes of rural 'development', deprivation, and dis/empowerment. *Sangtin Yatra* received an overwhelmingly positive response from progressive intellectuals and activists as well as a severe backlash from the prominent NGO where seven of the authors were employed. An analysis of these contrasting responses to *Sangtin Yatra* offered us another critical opportunity – this time with a country-wide and international readership in mind – to explore the controversial themes of NGO-ization and global feminisms, while also suggesting new possibilities for (re)imagining transnational feminist interventions and 'globalization from below' (Sangtin Writers 2006).[2]

The aftermath of critique also set us on another trajectory. It revealed, in jarring ways, that critique is allowed only by some people and that a differential price must be paid by people located at different levels in the global hierarchy of knowledge, for claiming a space as valid knowledge producers. Individually, it pushed us to seek allies and connected us to supporters in multiple institutions whose socio-political concerns intersected with ours.

Institutionally, it marked the beginning of new relationships and exchanges with educational organizations such as the National Council of Educational Research and Training in New Delhi, publishers such as Zubaan Press and the University of Minnesota Press, and aid organizations such as the Association for India's Development and Oxfam. These new encounters, in turn, made us recognize that the structures and practices of elitism, casteism, and classism that we had highlighted were by no means confined to the NGO sector but were present in varying configurations in all institutional spaces, and reimagining and reconfiguring them required us to interrogate them and generate critical dialogues in all the sites where intellectual and political work is being carried out, including our collective (Singh and Nagar 2006).

Within the Sangtin Collective, there have been at least three levels of internal critical reflexivity. The first concerns the roles, social locations, and privileges enjoyed by Richa Nagar and Richa Singh, and how these shape the politics of skills and labour in the group. The second pertains to the extent to which members of the collective are un/able to combat their own casteist, communalistic, and heterosexist values, and how this affects the relationships among us. The third addresses how the imagination of the collective's members is constrained by the same frameworks of donor-sponsored empowerment projects that we identified in *Sangtin Yatra* as NGO-ization of grassroots politics.

To grapple with these challenges, six members of the Sangtin team undertook another collective journey in August 2005 to explore how three different organizations in western India were negotiating their relationships among livelihoods of organization workers, pressures faced by donor-funded NGOs, and the need to build self-sustaining people's movements with a local base. The journey began by visiting Utthan, an NGO with a solid international donor-base, in Ahmedabad (Gujarat), where we learned about how that organization was trying to procure drinking water in areas with brackish waters, through the construction of tanks, check-dams and self-help groups. Our second set of meetings involved members of Self Employed Women's Association (SEWA), Ahmedabad, where we tried to understand their development of a highly successful women's dairy co-operative in the area. We learned about SEWA's evolution within a trade-union activism framework, its membership and working structures, and their limitations, especially in the realm of decision-making regarding funding and their ability to address questions of communal and patriarchal violence. Our last visit was with Mazdoor Kisan Shakti Sangathan in Devdoongri (Rajasthan), where we explored the renegotiation of its relationships with NGOs, donors' agencies, and NRI (non-resident Indian) support; and how it confronted questions of hierarchy within itself on an ongoing basis and created a broad, self-sustaining base of local peasants and workers to keep the movement dynamic and growing.

The trip triggered debate in our group with respect to the strengths and limitations of each organization as well as the future path(s) that Sangtin should embrace. The dialogues centred on the following. First, how to build a social movement based on local priorities and resources, rather than the visions and terms of specific donor groups. Second, although none of the organizations we visited focused explicitly on the politics of family, sexuality, or violence against women, our efforts to link each organization's approach towards casteism, communalism, and patriarchy with Sangtin's work sparked tough conversations on intimacies across Sawarn–Dalit and Hindu–Muslim divides, and on extra-marital relationships between women and men. Rather than forcing a consensus, we recognized the need to actively create an open membership and dynamic structure so that collective understandings about intersections among developmentalism, patriarchy, and casteist and communal untouchability/endogamy could broaden and deepen with the active involvement of men (and more women) from local Dalit and Sunni communities.

Last, but not least, we discussed our continuing transnational alliance; its strengths, problems, and contradictions. We noted that attending to the politics of privilege and skills within a collective differentiated by class, caste and religious affiliations as well as by material and symbolic resources requires constant challenges to the *status quo* within the collective. It is only possible to nurture collective leadership by creating more equitable conditions for all members to participate and claim spaces in national/transnational intellectual and political conversations. Consequently, we made two new commitments: a social agitation to release the waters of a major canal that have been siphoned away from the lands of the poorest peasants in 40 villages of Sitapur; and the launch of Sangtin's newspaper, *Hamara Safar*, as a forum for self-critique and analysis within the collective, and an opportunity for external dialogue and critique.

Beyond the impasse?

Dominant definitions of academic research automatically label as 'extra-curricular' intellectual production that falls beyond the sphere of certified university scholarship. Yet, without enduring connections with specific political struggles and critical dialogues with those whose interests progressive research seeks to serve, the 'sole-conceptualized'/'sole-authored' model of research will have little use-value for those who live in the 'fields' of the academics and NGO workers. Challenging normative definitions of knowledge and knowledge producers inevitably entails the reflexive reshaping of meanings embedded in socio-political contexts, languages, and institutional cultures and struggles.

The difficulties of 'speaking with' collaborators across borders are rooted in the structural realities of fieldwork. For feminist researchers who travel back and forth between structurally unequal worlds, this aspect of field-work has to be replaced by innovative and dynamic processes of collective knowledge production that are valued (as empowering/socio-politically pertinent) by those in the 'field' with whom we share political commit-ments. Indeed, in our view, the dangers of relinquishing responsibility for acquiring, producing, and disseminating knowledge about and by people inhabiting the rest of the world have never been greater. To leave total con-trol of what is said, and therefore widely 'understood' (whether about India, Tanzania, Iraq, or Palestine) to those whose interests lie within the spheres of global capital, an increasingly homogenized media, or the political *status quo* (as long as it serves US purposes) is to facilitate those very interests. Nor is it enough to simply criticize these processes and interests, and discuss them among ourselves – favourite academic pastimes.

Processual reflexivity and crossing borders with situated solidarities require openness to rethinking dominant standards of academic productiv-ity. Orchestrating such a shift entails challenging traditional academic norms that inhibit collective and collaborative research; privilege national and/or Western theorizing; and caution graduate students and early career academics against intellectual interests that coalesce around political con-cerns, issues, people, and modes of analysis that challenge institutionalized ways of knowing and thinking. Many Feminist and Women's Studies departments and programmes have celebrated 25-year anniversaries, sig-nalling a generation of scholarship within the academy. Feminist scholars can become important change agents within academic institutions and transdisciplinary feminist scholarship can facilitate collective and collabora-tive international research, and take the lead in redefining how we value community-based 'service'. It can be critical in moving us beyond the impasse by creating institutional spaces promoting a far broader view of what counts as significant scholarship, and by encouraging graduate students and faculty to take intellectual and political risks, including that highly charged and thoroughly scrutinized practice known as feminist fieldwork.

Acknowledgements

Initially authored by Richa and Susan between 1997 and 2000, this chapter was revised by Richa with the support of the Institute of Global Studies and the Office of International Programs at the University of Minnesota, a University of Minnesota CLA Research Fellowship Supplement, and a fellowship from the Center for Advanced Study in the Behavioral Sciences at Stanford. We have benefited tremendously from David Faust's sustained critical engagement, and from invaluable feedback provided by Lisa Disch, Mona Domosh, Sharilyn Geistfeld, M. J. Maynes, Jennifer Pierce, Naomi Scheman, Eric Sheppard and Adam Tickell.

NOTES

1 We recognize the problematic nature of these homogenizing dichotomies. Here, we use them to refer to an unequal structure of knowledge production – rooted in post/colonial hierarchies – where scholars in rich institutions of the North continue to dominate the international context in which knowledge about southern peoples and places is produced, circulated and discussed.

2 This book, *Playing with Fire*, was published simultaneously by the feminist press, Zubaan, in New Delhi and by the University of Minnesota Press. Zubaan embraced all the nine authors as 'Sangtin Writers', but market and cataloging considerations meant that Minnesota identified the authors as Sangtin Writers and Richa Nagar (2006).

23 RESEARCHING HYBRIDITY THROUGH 'CHINESE' BUSINESS NETWORKS

Henry Wai-chung Yeung

Reflecting on the culturalist perspective of Chinese capitalism

My aim is to reflect critically on my own posititonality in researching hybridity in Chinese business networks. The chapter is intentionally reflexive and thus peculiar to my own experience of researching Chinese capitalism. Chinese capitalism is defined as a historically- and geographically-specific form of economic organization among ethnic Chinese living outside mainland China, particularly in East and Southeast Asia. In studying Chinese capitalism, I have increasingly shifted from a static culturalist perspective to a dynamic transformative one, which, as I will argue, is a result of both critical reflections on my own research positionality as well as the material processes that increasingly hybridize Chinese capitalism.

Over the past ten years, some of the texts I have written about Chinese capitalism have shaped, albeit in a small way, the object I have studied. Exchanges during my interviews with key actors in Chinese family firms, or in academic workshops and conferences, or through readership of my published work, have produced the unintended effect of altering Chinese capitalism.

When I first began the research, I was naïve enough to believe in the intrinsic importance of *guanxi* or social relationships in explaining the economic organization of Chinese capitalism. I saw the Chinese cultural trait of *guanxi* as *essentially* shaping the behaviour and activities of ethnic Chinese business firms. While my fascination with culture and ethnicity in shaping economic organizations predates the 'cultural turn' in economic geography (Lee and Wills 1997; Thrift and Olds 1996), I did not realize that I was substituting one form of essentialism for another – *cultural essentialism* for *economic essentialism*. In retrospect, my cultural essentialism had much to do with my positionality as an ethnic Chinese born in mainland China and as a citizen in a predominantly ethnic Chinese economy (Singapore), and a permanent resident in another under the authority of the People's

Republic of China (Hong Kong SAR). I thought I was gifted with 'insider' knowledge. But as the famous Chinese saying goes, an insider is so obsessed with a phenomenon that it takes an outsider to tell a better picture.

When I began my research about ethnic Chinese entrepreneurs from Hong Kong in 1993, I was attracted to popular writings about Chinese capitalism. In retrospect, their depictions were nothing more than ethnic stereotypes or, worst, racist caricatures. In this vast literature, Chinese culture and *guanxi* networks were celebrated as the fundamental reason for the success of leading Chinese entrepreneurs and their business conglomerates in the Asia Pacific region. The popularity and prominence of such a triumphant characterization of Chinese capitalism as 'bamboo networks', '*guanxi* capitalism' or 'Confucian capitalism' perpetuated an essentialist argument that all ethnic Chinese in East and Southeast Asia share similar Confucian worldviews. In this 'culturalist' literature, the Chinese Confucian culture was taken for granted as both presupposing the economic success of 'Overseas Chinese' and defining the 'spirit of Chinese capitalism'. Defined as an enduring cultural system of economic activities, Chinese capitalism was written as static and self-contained. Prompted by such celebratory readings, much of my early research was concerned with unpacking Chinese business activities from a culturalist position. Clearly, I fell prey to the triumphant and dominant discourse of Chinese capitalism that permeated much of East Asian studies at the time.

During the same period, economic geography took up the idea of 'social embeddedness'. Popularized by the economic sociologist Mark Granovetter (1985), embeddedness referred to the intertwined relationships between economic action and social structures. It became a key concept in refiguring the economic in economic geography (Dicken and Thrift 1992; Thrift and Olds 1996). While it represented a move away from neoclassical economics, social embeddedness implied a form of 'structuralist' thinking because economic processes are understood as inseparable from and thus embedded in pre-existing structures of social and cultural networks. These pre-existing social structures are viewed as durable and persistent.

At the time, it seemed to make sense for me to conceive Chinese business as firms socially embedded in dense networks of *guanxi* or personal relationships. More crucially, I was so determined methodologically to uncover the 'embeddedness variable' in my early empirical study of Hong Kong investments in Southeast Asia (Yeung 1997) that I failed to notice the important underlying transformations and *disembedding processes* taking place in the socio–economic (re)organization of Chinese capitalism. Instead, I went into the field with the firm belief that Chinese capitalism was about ethnic Chinese and their cultural beliefs that drive the economic system.

At both theoretical and methodological levels, I paid no attention to the possibilities of *non-Chinese* (e.g., foreign partners, institutional investors and international media commentators) as important actors in shaping the nature and organization of Chinese capitalism. Rather, I was driven by a culturalist perspective, attuned to discovering cultural convergence and structural relations of embeddedness rather than dynamic global change and transformations. The end result of this self-imposed 'lock-in' trajectory in my research practice were texts that celebrated the importance of *guanxi* and personal relationships in explaining the transnational business networks of Chinese business firms from Hong Kong (Yeung 1997, 1998). Methodologically, I saw only what I wanted to see, failing to shed light on perhaps the even more important story – the dynamic transformation of Chinese capitalism in a global era.

Repositioning my research: what has culture got to do with it?

By the late 1990s I became increasingly aware of my own research positionality. I realized that I was more than just an academic researcher who produces knowledge on Chinese capitalism. My research 'findings' were impacting the very subjects of the research. A special report on Asian family succession by John Ridding, a Hong Kong-based correspondent of *The Financial Times*, on 10 March 1998 quoted me: 'For many international investors the lack of transparency is linked to family control and succession' (p. 15). In turn, this opened many doors. The *Far Eastern Economic Review* subsequently approached me for an exclusive interview on 'succession in Chinese family business' that was also aired live on CNBC Live shortly after the FT report.

The coverage of my views made me aware for the first time that academics could be important in shaping the business community. I was sure that the interest in my comments was heightened by the ongoing accusation of 'crony capitalism' in the wake of the Asian economic crisis. Such leading commentators of the Asian economic crisis in 1997/1998 as Paul Krugman and Alan Greenspan argued that the crisis resulted from the lack of transparency in the financial systems of various Asian economies. In East and Southeast Asia, the strong patron–client relationships between the state and business people (mostly ethnic Chinese), and rooted in Confucianism, was posited as the primary cause of the Asian economic crisis. My suggestion of a link between family control and lack of transparency in Chinese family firms couldn't have been timelier!

It was clear given the resulting media interest, though, that I needed to question my own positionality in this project about Chinese capitalism. It was producing unintended effects through media and professional publications. As an ethnic Chinese researcher speaking on the perils of 'Chinese capitalism', my words were likely to be construed and received differently from those of 'outsiders' such as Anglo-American scholars. I wondered, in fact, whether I had become part of the alleged 'Western conspiracy' that, according to the former Prime Minister of Malaysia, Dr Mahathir Mohamed, brought the crisis to Asia. Consequently, I began questioning my own ethnic and cultural identity as an 'Overseas Chinese' – overseas to where and what? Am I a believer in Confucianism and *guanxi* capitalism? Do I know and practice *guanxi* as all ethnic Chinese are allegedly expected to do? And what about the Italians and the Jews or Anglos, don't their forms of economic activity rest on embedded relations even though it is not called '*guanxi*'? In short, I began to lose faith in the culturalist perspective, and began searching for an alternative approach to study Chinese capitalism.

Why was there this methodological u-turn? During the course of this decade-long research into Chinese capitalism, what changed was not the subject of my intellectual pursuit, but rather my thinking about it. My reasoning became less culturalist, more transformative and reflexive. While the change in my views was partly an internal response, it was also correlated with the dramatic pace at which globalization tendencies externally transformed the world economy, integrating Chinese capitalism within it (as suppliers, strategic partners, and competitors). It is this story of *change* and *transformation* in Chinese capitalism that underscores much of the concept *hybridity*. Chinese capitalism does not exist as a 'pure' form of capitalism. It evolves over time, and forms *hybrid capitalism* defined by its incomplete, partial and contingent transformations. We are now witnessing the rise of this peculiar hybrid. It has no definite end-state and form because political and economic outcomes of transition depend on the continuing complex interactions among social actors and their embedded institutions. Moreover, the very territorial unboundedness of Chinese capitalism contributes to its emergent hybridity given the supra-national multiple forces that condition its structures and trajectory. Chinese capitalism is now a curious mixture and juxtaposition of traditional practices and established global norms and conventions (Yeung 2004).

Chinese capitalism, then, fits Hamilton's (1999: 4) thesis that:

> capitalism is not a stable and readily identifiable configuration that, like a flower, suddenly bursts forth in bloom. Instead, capitalism is merely a term that covers an extremely wide range of diverse economic activities organized in the context of competitive markets and whose institutional conditions include private ownership and non-state decision making.

Likewise, it is congruent with Storper and Salais' (1997) idea of the economy as a hybrid object, constituted through a diversity of conventions and worlds of production in which actors organize and legitimize their action and behaviour. As a result, the concept of culture in Chinese capitalism needs to be reconceptualized as a repertoire of historically contingent and geographically specific practices that respond and adapt to changing local, regional and global circumstances rather than as permanently fixed mental and organizational structures that resist challenges and pressures to change.

This dynamic approach is necessary because Chinese capitalism is not a cultural artefact, forever cast in stone. Rather, Chinese capitalism is *lived* as a peculiar set of spatial-temporal outcomes, is highly institutionalized, and consists of immense struggle and contestation by *social actors* through discursive strategies and material practices (of which I played a small role through my research and writings). I have realized it is important to focus on key social actors, rather than a focus on the cultural systems of beliefs and values in Chinese capitalism. Taking such an *actor-oriented approach*, however, should not throw 'culture' out with the popular writings on the 'Overseas Chinese business'. Instead, it is about seeing culture as a set of practices that change with circumstances and contexts, *and* allowing us to see Chinese capitalism as a hybrid form.

Researching into the secrets of Chinese capitalism

Let me now turn to the politics and practice of my research during my decade-long project on Chinese capitalism. Elsewhere (Yeung 2003), I discussed a process-based methodological framework for understanding the new economic geographies. Figure 23.1 provides a convenient summary. While the following *post-hoc* rationalization of my research framed within Figure 23.1 might appear too neat and clean, the realities are much messier and *ad hoc*. Methodological pragmatism and improvisation are my rule of thumb.

In Chinese capitalism, not only are economic institutions embedded in social relations, but actors in these networks of relations also have hybrid identities that are strongly influenced by their discursive and institutional contexts. Figure 23.1 shows that an integrated method approach is an appropriate methodology to unfold the secrets of Chinese capitalism. Starting with the *choice of empirical data*, I knew at the beginning of my research that an exclusive reliance on either primary data or secondary sources would be insufficient to unlock the secrets of Chinese capitalism. Instead, I had to combine both data sources and triangulate their efficacy and complementarity. For example, conducting large-scale surveys with such actors in Chinese capitalism as CEOs of family firms – a methodological approach much favoured in business school research (Ahlstrom et al. 2004; Tsang 2002) – may be sufficient to generate some quantifiable

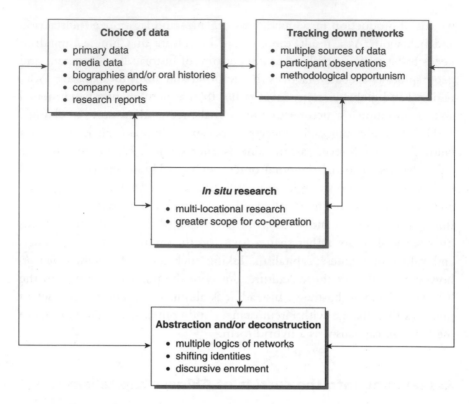

FIGURE 23.1   The integrated method approach to Chinese capitalism research

establishment characteristics (e.g., firm size and employment) and statistical relationships on certain dependent variables (e.g., family ownership and financial performance), but it is not a meaningful method to explain the material and discursive constructions of Chinese family firms and their business networks, let alone provide deep insights into their transformations in relation to globalization tendencies. In-depth personal interviews are much more relevant.

To this end, I conducted two major rounds of personal interviews with top executives from 73 Hong Kong-based Chinese family firms in 1994 and 54 Singapore-based Chinese family firms in 1998 (Yeung 2004: Chapter 4). During these rounds of data collection, I focused on both firm-level data and some highly personalized information (e.g., on *guanxi* connections and business network formation). This process of interview-based data collection, however, does not negate the relevance of quantitative data. In fact, I adopted an unconventional approach to these personal interviews (Yeung 2002). I conducted those interviews with conventional questionnaires designed to extract quantifiable data on causal factors and variables. During each interview, I asked questions sequentially as listed on the questionnaire – a common experience in all questionnaire surveys. Unlike most questionnaire surveys, however, the interviewees were fully able to offer detailed comments and further elaborations on issues of their interest. This qualitative information was captured through tape-recording with prior consent. Meanwhile, I also probed further their answers to the questionnaires that might be important in generating unique theoretical insights. As such, my empirical data collection method was neither entirely qualitative nor a large-scale impersonal survey. Rather, it was a curious hybrid of both quantitative surveys and qualitative personal interviews – what I prefer to call 'interview surveys'. At the end of each interview, I would have one completed questionnaire and, in most cases, at least one hour-long tape-recording that contained good qualitative information. This methodological pragmatism enabled me to compile a large quantifiable database on Chinese family firms in Hong Kong and Singapore and a large collection of interesting vignettes that could be further developed into qualitative case studies. In this way, I managed to bridge the quantitative–qualitative divide in existing research of Chinese capitalism by collecting both kinds of data in a single interview survey. This procedure thus minimized the usual methodological problem of collecting large amount of quantitative data through highly impersonal postal surveys in which we actually have no idea *who* completed the questionnaire. It also circumvented the messy problem of analyzing qualitative data collected from open-ended personal interviews (Alvesson 2003). My interview surveys therefore allowed for a combination of quantitative and qualitative information from the *same* interviewees.

Meanwhile, a large amount of secondary data exists in the form of media interviews, biographies, oral histories, company documents, and research reports (see Figure 23.1). These secondary data can sometimes be very useful as a starting point to identify different kinds of actors in 'Chinese' business networks. For example, the media often have special access to many leading actors in Chinese capitalism because they are well connected and willing to talk to reporters. These elite actors may be less likely to talk to researchers who, as McDowell (1998) and Herod (1999) have observed, have little to reciprocate. Well-known credit-rating agencies and international business media often have very good access to large Chinese business firms. For example, in Olds and Yeung (1999), we used a confidential company report from a credit-rating agency to analyze the proposed corporate reorganization of the largest Chinese business conglomerate from Hong Kong – Li Ka-shing's Cheung Kong Group.

Other financial documents were used in Yeung (2004) to showcase the significant transformations in the social organization of Chinese capitalism towards hybrid capitalism. In this case, Li Ka-shing made very good use of enrolling financial analysts into his actor networks. The enrolment of non-Chinese elites (financial analysts) into 'Chinese' business networks represents a pragmatic strategy for leading Chinese actors to capitalize on globalization tendencies (financial globalization in this case). Biographies and/or oral histories of selected actors in Chinese capitalism may also offer a significant depth of information unavailable to most researchers (e.g., Chan and Chiang 1994; Tsui-Auch 2004). By definition, these secondary data were collected for different reasons and might contain inherent bias. I was convinced that *both* primary and secondary data were needed to satisfy the criteria of data triangulation. By triangulation, I mean the deployment of different research methods and thus different data to cross-verify and complement each other in an actual research process. In my project, I used personal interviews to verify some of the data and information contained in media and other unpublished reports. I also used secondary data to fill the gaps in primary data collected from interviews with leading actors in Chinese capitalism.

Having identified the main actors in Chinese capitalism, it is now possible to *trace the interconnections* of actors in 'Chinese' business networks (see Figure 23.1). My earlier discussion of the socio-spatial attributes of these actors does not imply that they must be all powerful. Power is a relational construct; it 'is not some "thing" that moves, but an effect that is mediated, and such effects may mutate through relations of successive or simultaneous reach' (Allen 2003: 37). An actor's power is not a given attribute because its efficacy can only be known when it is realized. For example, a patriarch in the family business may not be as powerful as we imagine if there is a significant contestation of power among the siblings outside the control of this patriarch. Tracking down actor networks in Chinese capitalism is theoretically and methodologically challenging.

To begin this tracking, a good combination of multiple methods and data is critical to identifying the key actors involved in these networks. Participant observation is a valid and reliable method to 'follow through' these networks and their strategic outcomes. The reality of empirical research, however, is often not as neat as proposed in methodological 'cookbooks'. In my own research, I had to terminate prematurely my participant observation precisely because my identity as an academic researcher became too difficult for those business actors to handle. Some actors in the network did not want to be included in the study, fearing their personal information might be made public through my project writings. Here methodological opportunism emerges as an appropriate guideline in the sense that we need to be highly receptive to and prepare for new opportunities that may open doors for us. As argued by Ward and Jones (1999) in their reflections on researching local elites in England, the political-temporal contingency of the research process and the politically time-specific entry made by the researcher into the research field can significantly shape the interaction between the researcher and the researched. In my study on Hong Kong firms, for example, I capitalized on my *guanxi* or personal relationships with the family members of one ethnic Chinese entrepreneur to arrange for in depth interviews. He was subsequently able to reach several other key actors of the quasi-family business network in Hong Kong and in Singapore.

Tracing actor networks clearly has a spatial dimension and *in situ* research in each local context is an indispensable method to go beyond the 'remote sensing' approach to studying today's complex social and economic life (see Figure 23.1). By 'remote sensing', I mean researchers who study Chinese capitalism from a distance through postal questionnaire surveys and the like. Because many actor networks in Chinese capitalism are multinational, however, it is imperative to engage in multi-locational fieldwork to track down these networks. Many such actor networks are increasingly transcending Asia in their geographical scope (e.g., Olds 2001; Zhou and Tseng 2001). Conducting multi-locational *in situ* research, though, creates serious time and resource constraints. But transnational research offers many benefits, including deeper insights into the material and discursive constructions of actor networks and a better understanding of spatially differentiated contexts of their constructions.

This multi-locational and multidisciplinary research into Chinese capitalism takes me to the most important aspect of the integrated method approach – abstraction and deconstruction (see Figure 23.1). Instead of being overtly concerned with describing everything about actor networks in Chinese capitalism – a methodological procedure often found in ethnography (e.g., Yao 2002) and some quarters of actor network theory (e.g., Murdoch 1997) – I was interested in unfolding the multiple logic(s) and mechanisms of these actor networks. This approach to Chinese capitalism requires me to focus on key social actors and their dynamics in different spatial and temporal settings. I believed that these processes do not emerge directly from empirical observations. In other words, data do not speak for themselves. I needed abstraction to distill these multiple logic(s) and causal mechanisms of Chinese capitalism from an array of messy empirical data, and to facilitate theory development (see Orrù et al. 1997; Whitley 1999). Equally important, I wanted to go beyond a materialist theory of Chinese capitalism that focuses narrowly on economic structures and that is oblivious to its discursive constructions. Deconstruction in the form of discourse analysis is therefore a useful method in order to identify different discourses that underpin the rise of Chinese capitalism and Chinese business networks. By decentring hegemonic discourses of Chinese capitalism as essential products of Confucian culture, I sought to theorize the time-space dynamism of everyday life by Chinese and non-Chinese actors and bound up with specific hybrid networks of social and economic relationships.

Implications for politics and practice in economic geography

I would like to finish by discussing several implications that stem from what I have written about how we might think about politics and practice in economic geography. While it is almost axiomatic to say that research is *not* a neutral act, I think it remains imperative for practising economic geographers to recognize their own *positionality* in any research projects (Kong 2004; Yeung and Lin 2003). Had I been more aware of my positionality early on, it would have saved many pages of journal space. I should have been more aware of how my own positionality shaped my 'over-socialized' and over-embedded view of Chinese capitalism. After all, my perceived advantage as an ethnic Chinese turned out not to be advantageous at all – I was 'locked in' to my culturalist worldview. If recognizing one's positionality is paramount to a reflexive research process, not falling victim to established methodological procedures is another. I should not and did not have to start my research into Chinese capitalism with business firms, even though economic geography at the time was dominated by a concern with industrial firms. With hindsight, I could have taken a different methodological route, adopting, for example, a qualitative social network analysis in the spirit of actor network theory. I could have delved into the hybrid identities of second- and third-generation Chinese entrepreneurs, most of whom received their tertiary education in English-speaking countries and have tremendous international business experience. And the wish list goes on.

My point is that economic geographers should be open to *methodological experimentation*; we should not stay within our methodological 'comfort zones' as quantitative modellers did in the 1960s or the practitioners of seemingly endless case studies who dominate economic geography today. If economic geographers can be pluralistic and adventurous in their flirtations with social theories, why not spare a little time and effort to strengthen the discipline's methodological foundations? All too often, our 'philosophical horse' does not come with a 'methodological cart'. We are leaving too much of our methodology and practice to other social scientists like sociologists. While research methods and practices are highly guarded territory in sociology, with its specialized journals (*Sociological Methods and Research*, *Sociological Methodology* and *Quality and Quantity*), methodological rigour is much less an issue in human geography, let alone economic geography.

This methodological experimentation that I urge requires a significant mindset change. First, it necessitates our role as active academicians. We need to interact with our research 'subjects', not in the form of an impersonal encounter between the researcher and the researched. We need to understand that the research act is not just about learning from our 'subjects', but it is also about them gaining insights into our geographical imaginations. As an example, I very much enjoy the few minutes of 'lecturing' CEOs about what (economic) geography is whenever they ask why am I coming from the Department of Geography. A mundane research interview might just turn out to be a highly productive two-way learning process. Second, we have to be aware of the inherent fallibility of our knowledge, just like anybody else. Contrary to conventional wisdoms, we do not have privileged access to the world around us. Social actors have as important and subjective knowledge about the fascinating workings of the economic-geographical worlds as we do. So where does that leave us in this knowledge production system? We must drop our esteemed claim to truth and begin to work on 'local knowledges'. By local knowledges, I mean highly contextualized understanding of patterns and processes in particular geographical settings. These local knowledges might just come from ordinary people rather than academics. We need an economic geography that is methodologically more attuned to 'local knowledges'. Yet we must always remain reflexive, not becoming engrossed in a single paradigmatic explanation as I was by using culture to explain the spirit of Chinese capitalism.

Acknowledgements

I would like to thank Markku Tykkyläinen and the Geographical Society of Finland for inviting me to their Annual Meeting in 2004 where this paper was first given. Trevor Barnes offered insightful comments on different drafts of this chapter. I am solely responsible for its content.

References

Agamben, G. (1993) *The Coming Community*, Minneapolis: University of Minnesota Press.

Agarwal, B. (1982) *Agricultural Modernisation and Third World Women*, Geneva: International Labour Organisation.

Aglietta, M. (1987 [1979]) *A Theory of Capitalist Regulation*, London: Verso.

Ahlstrom, D., Young, M. N., Chan, E. S. and Bruton, G. D. (2004) 'Facing constraints to growth?', *Asia Pacific Journal of Management* 213: 263–85.

Allen, J. (2003) *Lost Geographies of Power*, Oxford: Blackwell.

Althusser, L. (1971) *Lenin and Philosophy and Other Essays*, New York: Monthly Review Press.

Alvesson, M. (2003) 'Beyond neopositivists, romantics, and localists', *Academy of Management Review* 28: 13–33.

Amin, A. and Thrift, N. (2004) *The Blackwell Cultural Economy Reader*, Oxford: Blackwell.

Anderson, E. (2004) 'Uses of value judgements in science: a general argument, with lessons for a case study of feminist research on divorce', *Hypatia* 19: 1–24.

Antipode (2001) 'Commemorating the life and work of Bennett Harrison', *Antipode* 33: 17–71.

Appadurai, A. (1988) 'Place and voice in anthropological theory', *Cultural Anthropology* 31: 16–20.

Arthur, B. (1994) *Increasing Returns and Path Dependence in the Economy*, Ann Arbor: University of Michigan Press.

Babcock-Lumish, T. (2005) 'Venture capital decision-making and the cultures of risk', *Competition and Change* 9: 329–56.

Bagguley, P., Lawson, M., Warde, M., Shapiro, D. and Urry, J. (1990) *Restructuring: Place, Class and Gender*, London: Sage.

Bailey, A., Wright, R., Mountz, A. and Miyares, I. (2002) '(Re)producing Salvadoran transnational geographies', *Annals, Association of American Geographers* 92: 125–44.

Baldwin, J. (1995) *The Dynamics of Industrial Competition*, Cambridge: Cambridge University Press.

Banks, M. and MacKian, S. (2000) 'Jump in! The water's warm: a comment on Peck's Grey Geography', *Transactions, Institute of British Geographers* 25: 249–54.

Barnes, T. J. (1987) 'Homo economicus, physical metaphors, and universal models in economic geography', *The Canadian Geographer* 31: 299–308.

Barnes, T. J. (1996) *Logics of Dislocation*, New York: Guilford Press.

Barnes, T. J. (2001) 'Retheorizing economic geography: From the quantitative revolution to the "cultural turn"', *Annals of the Association of American Geographers* 91(3): 546–65.

Bartkevicius, J. n.d. 'Forum on nonfiction: literal versus invested truth in memoir', *Fourth Genre* msupress.msu.edu/journals/fg/.

Begg, R. and Pickles, J. (1998) 'Institutions, Social Networks, and Ethnicity in the Cultures of Transition: Industrial Change, Mass Unemployment and Regional Transformation in Bulgaria'; in J. Pickles and A. Smith (eds) *Theorising Transition: The Political Economy of Post-Communist Transformations*. London: Routledge.

Behar, R. (1996) *The Vulnerable Observer*, Boston: Beacon Press.

Beneria, L. and Sen, G. (1982) 'Class and gender inequalities and women's role in economic development: Boserup revisited', *Feminist Studies* 8: 157–76.

Benton, S. (2004) 'Critical intellectuals in the post-socialist world', *Soundings* 27: 10–18.

Bergman, B. (1990) 'Feminism and economics', *Women's Studies Quarterly* 18: 68–74.

Bernstein, H. (1979) 'African peasantries: a theoretical framework', *Journal of Peasant Studies* 6: 421–43.

Berry, S. (1984) 'The food crisis and agrarian change in Africa', *African Studies Review* 27(2): 59–112.

Bertrand, M., Karlan, D., Mullainathan, S., Shafir, S. and Zinman, J. (2005) 'What's psychology worth? A field experiment in the consumer credit market', *Working Paper 11892*, Cambridge: National Bureau of Economic Research.

Beynon, H., Hudson, R. and Sadler, D. (1994) *A Place Called Teesside*, Edinburgh: Edinburgh University Press.

Blaikie, P. and Brookfield, H. (1987) *Land Degradation and Society*, New York: Methuen.

Blomley, N. (1994) 'Activism and the academy', *Environment & Planning D: Society & Space* 12: 383–5.

Blomley, N. (1995) 'Reply to Tickell', *Environment & Planning D: Society & Space* 13: 239–40.

Bluestone, B. and Harrison, B. (1982) *The Deindustrialization of America*, New York: Basic Books.

Boal, A. (1998) *Legislative Theatre: Using Performance to Make Politics*, New York: Routledge.

Böröcz, J. (2000) 'The Fox and the Raven: The European Union and Hungary Renegotiate the Margins of "Europe"', *Comparative Studies in Society and History* 42: 847–75.

Bourdieu, P. (1977) *Outline of a Theory of Practice*, Cambridge: Cambridge University Press.

Bourgois, P. (1996) 'Confronting anthropology, education, and inner-city apartheid', *American Anthropologist* 98: 249–65.

Bowlby, S., Lewis, J., McDowell, L. and Foord, J. (1989) 'The geography of gender', in R. Peet and N. J. Thrift (eds) *New Models in Geography* (Vol. 1), London: Unwin Hyman.

Boyer, R. (1990) *The Regulation School: A Critical Introduction*, New York: Columbia University Press.

Brakman, S., Garretsen, H. and van Marrewijk, C. (2001) *An Introduction to Geographical Economics*, Cambridge: Cambridge University Press.

Braverman, H. (1974) *Labor and Monopoly Capital*, New York: Monthly Review Press.

Brenner, N., Jessop, B., Jones, M. and MacLeod, G. (eds) (2003) *State/Space: A Reader*, Oxford: Blackwell.

Bryant, R. L. (1992) 'Political ecology: an emerging research agenda in Third-World Studies', *Political Geography* 11: 12–36.

Burawoy, M. (1979) *Manufacturing Consent*, Chicago: University of Chicago Press.

Burawoy, M. (1991) 'The extended case method', in M. Burawoy, A. Burton, A. A. Ferguson and K. J. Fox (eds) *Ethnography Unbound*, Berkeley: University of California Press.

Burawoy, M., Blum, J. A., George, S., Gille, Z., Gowan, T., Haney, L., Klawiter, M., Lopez, S., Riain, S. O. and Thayer, M. (2000) *Global Ethnography*, Berkeley: University of California Press.

Butler, J. (1993) *Bodies that Matter*, New York: Routledge.

Caerlewy-Smith, E., Clark, G. and Marshall, J. (2006) 'Comment. Agitation, resistance, and reconciliation with respect to socially responsible investment', *Environment & Planning A* 38: 1585–9.

Cameron, J. and Gibson-Graham, J. K. (2003) 'Feminising the economy', *Gender, Place and Culture* 10: 145–57.

Carney, J. (1993) 'Converting the wetlands, engendering the environment: the intersection of gender with agrarian change in The Gambia', *Economic Geography* 69: 329–49.

Carney, J. A. (2001) *Black Rice: The African Origins of Rice Cultivation in the Americas*, Cambridge, MA: Harvard.

Carney, J. (2003) 'African traditional plant knowledge in the circum-Caribbean region', *Journal of Ethnobiology* 23: 167–85.

Carney, J. (2004) '"With grains in her hair": rice history and memory in colonial Brazil', *Slavery and Abolition* 25: 1–27.

Carney, J. (2005) 'Rice and memory in the age of enslavement', *Slavery and Abolition* 26: 325–48.

Carney, J. and Voeks, H. (2003) 'Landscape legacies of the African diaspora in Brazil', *Progress in Human Geography* 27: 139–52.

Castree, N. (2000) 'Professionalism, activism and the university: wither "critical geography"?' *Environment & Planning A* 32: 955–70.

Castree N. (2004) 'Economy and culture are dead! Long live economy and culture!', *Progress in Human Geography* 28: 204–26.

Chakrabarty, D. (2000) *Provincializing Europe: Postcolonial Thought and Historical Difference*, Princeton: Princeton University Press.

Chan, K. B. and Chiang, S.-N. C. (1994) *Stepping Out: The Making of Chinese Entrepreneurs*, Singapore: Simon and Schuster.

Charoenloet, V. (1989) 'The crisis of state enterprises in Thailand', *Journal of Contemporary Asia* 19: 206–17.

Charoenloet, V. (1998) 'The situation of health and safety in Thailand', Unpublished manuscript: Faculty of Economics, Chulalongkorn University.

Chouinard, V. and Fincher, R. (1983) 'A critique of structural Marxism and human geography', *Annals, Association of American Geographers* 73: 137–50.

Christopherson, S. (2001) 'Bennett Harrison's gift: collaborative approaches to regional development', *Antipode* 33: 29–33.

Clark, G. L. (1998) 'Stylised facts and close dialogue: methodology in economic geography', *Annals, Association of American Geographers* 88: 73–87.

Clark, G. L. (2001) 'Vocabulary of the new Europe: code words for the millennium', *Environment & Planning D: Society & Space* 19: 697–717.

Clark, G. L. (2003) *European Pensions and Global Finance*, Oxford: Oxford University Press.

Clark, G. L. (2006) 'Setting the agenda: the geography of global finance', in S. Bagchi-Sen and H. Lawton Smith (eds) *Economic Geography: Past, Present and Future*, London: Routledge.

Clark, G. L., Caerlewy-Smith, E. and Marshall, J. (2006a) 'The consistency of UK pension fund trustee decision-making', *Journal of Pension Economics and Finance* 5: 91–110.

Clark, G. L., Caerlewy-Smith, E. and Marshall, J. (2006b) 'Solutions to the asset allocation problem by informed respondents: the significance of the size-of-bet and the 1/n heuristic' *WPG06-12*, Oxford: Oxford University Centre for the Environment.

Clark, G. L., Caerlewy-Smith, E. and Marshall, J. (2007) 'Pension fund trustee competence: decision making in problems relevant to investment practice', *Journal of Pension Economics and Finance* 6: 67–86.

Clark, G. L., Feldman, M. and Gertler, M. S. (eds) (2000) *The Oxford Handbook of Economic Geography*, Oxford: Oxford University Press.

Clark, G. L. and Wójcik, D. (2007) *The Geography of Finance*, Oxford: Oxford University Press.

Cloke, P. (2002) 'Deliver us from evil', *Progress in Human Geography* 26: 587–604.

Collier, A. (2003) *In Defence of Objectivity*, London: Routledge.

Community Economies Collective (2001) 'Imagining and enacting non-capitalist futures', *Socialist Review* 28(3+4): 93–135.

Confessore, N. (2002) 'Comparative advantage: how economist Paul Krugman became the most important political columnist in America', *Washington Monthly* 34: 19–26.

Connolly, W. (1995) *The Ethos of Pluralization*, Minneapolis: University of Minnesota Press.

Connolly, W. (1999) *Why I am Not a Secularist*, Minneapolis: University of Minnesota Press.

Conti, A. (1979) 'Capitalist organization of production through non-capitalist relations', *Review of African Political Economy* 15/16: 75–92.

Cook, I. (2001) '"You want to be careful you don't end up like Ian. He's all over the place": autobiography in/of an expanded field', in P. Moss (ed.) *Placing Autobiography in Geography*, Syracuse, NY: Syracuse University Press.

Cook, I. (2004) 'Follow the thing: papaya', *Antipode* 36: 642–64.

Cooke, P. (1986) 'The changing urban and regional system in the United Kingdom', *Regional Studies* 30: 243–51.

Corbridge, S. (1998) 'Reading David Harvey: entries, voices, loyalties', *Antipode* 30: 43–55.

Cornog, E. (2005) 'Let's blame the readers', *Columbia Journalism Review* 43(5): 43–50.

Crang, P. (1997) 'Cultural turns and the (re)constitution of economic geography: Introduction to section one', in R. Lee and J. Wills (eds) *Geographies of Economies*, 3–15. London: Arnold.

Cranmer, M. and McChesney, T. (2003) 'Chronic wasting disease: risks to hunters and consumers of deer and elk meat', *Neurotoxicology* 24: 73.

Cravey, A. J. (1998) *Women and Work in Mexico's Maquiladoras*, Lanham, MD: Rowman & Littlefield Inc.

Cravey, A. J. (2003) 'Toque una ranchera por favor', *Antipode* 35: 603–21.

Cravey, A. J. (2005a) 'Desire, work, and transnational identity', *Ethnography* 6: 357–83.

Cravey, A. J. (2005b) 'Working on the global assembly line', in L. Nelson and Seager, J. (eds) *A Companion to Feminist Geography*, Oxford: Blackwell.

Crowley, S. and Ost, D. (2001) (eds) *Workers After Workers' States: Labour and Politics in Postcommunist Eastern Europe*, Lanham: Rowman and Littlefield.

Czarniawka, B. (1998) *A Narrative Approach to Organization Studies*, London: Sage.

De Janvry, A. (1981) *The Agrarian Question and Reformism in Latin America*, Baltimore, MD: Johns Hopkins University Press.

De Vault, M. (1999) *Liberating Method: Feminism and Social Research*, Philadelphia: Temple.

Dean, J. (1995) *Solidarity of Strangers: Feminism after Identity Politics*. Berkeley: University of California Press.

Deere, C. D. (1976) 'Rural women's subsistence production in the capitalist periphery', *Review of Radical Political Economics* 8: 133–48.

Dey, J. (1981) 'Gambian women: unequal partners in rice development projects?', *Journal of Development Studies* 17: 109–22.

Dey, J. (1983) *Women in Rice Farming Systems with a Focus on Africa*, Rome: Food and Agricultural Organization.

Dezalay, Y. and Garth, B. (2002) *The Internationalization of Palace Wars*, Chicago: University of Chicago Press.

Diamond, D. (2004) *Practicing Democracy Final Report*, www.headlinestheatre.com.

Dicken, P., Kelly, P. F., Olds, K. and Yeung, HW-C. (2001) 'Chains and networks, territories and scales: towards a relational framework for analysing the global economy', *Global Networks* 1: 89–112.

Dicken, P. and Thrift, N. (1992) 'The organization of production and the production of organization', *Transactions, Institute of British Geographers* 17: 279–91.

Dorling, D. and Shaw, M. (2002) 'Geographies of the agenda', *Progress in Human Geography* 26: 629–46.

Downward, P. and Mearman, A. (2002) 'Critical realism and econometrics', *Metroeconomica* 53: 391–415.

Duncan, J. and Ley, D. (1982) 'Structural Marxism and human geography', *Annals, Association of American Geographers* 72: 30–59.

Dunn, E. C. (2003) 'Webs of resistance in a newly privatized Polish firm: workers react to organizational transformations', *British Journal of Industrial Relations* 41: 810–12.

Dunne, T., Roberts, M. and Samuelson, L. (1988) 'Patterns of firm entry and exit in US manufacturing industries', *Rand Journal of Economics* 19: 495–515.

Dupré, J. (2001) *Human Nature and the Limits of Science*, Oxford: Oxford University Press.

Duranti, A. (1992) 'Language and bodies in social space: Samoan ceremonial greetings', *American Anthropologist* 94: 657–91.

Durlauf, S. (2002) 'On the empirics of social capital', *Economic Journal* 112: 459–79.

Durlauf, S., Johnson, P. and Temple, J. (2004) 'Growth econometrics', in P. Aghion and S. Durlauf (eds) *Handbook of Economic Growth*, North-Holland: Elsevier.

Dyck, I. and McLaren, A.T. (2004) 'Telling it like it is? Constructing accounts of settlement with immigrant and refugee women in Canada', *Gender, Place and Culture* 11: 513–34.

Eliasoph, N. and Lichterman, P. (1999) '"We begin with our favorite theory ...": reconstructing the extended case method', *Sociological Theory* 17: 228–34.

Elson, D. (1979) 'The value theory of labour', in D. Elson (ed.) *Value: The Representation of Labour in Capitalism*, London: CSE Books.

England, K. (1993) 'Suburban pink collar ghettos: The spatial entrapment of women?', *Annals, Association of American Geographers* 83: 225–42.

England, K. and Stiell, B. (1997) '"They think you're as stupid as your English is": constructing foreign domestic workers in Toronto', *Environment & Planning A* 29: 195–215.

Espiritu, Y. L. (2003) *Home Bound: Filipino American Lives across Cultures, Communities, and Countries*, Berkeley: University of California Press.

Ettlinger, N. (2001) 'A relational perspective in economic geography: Connecting competitiveness with diversity and difference', *Antipode* 33: 216–27.

Fairclough, N. (1992) *Discourse and Social Change*, Cambridge: Polity Press.

Fairclough, N. (1995) *Media Discourse*, London: Edward Arnold.

Fairclough, N. (2003) *Analyzing Discourse*, London: Routledge.

Fama, E. and French, K. (2005) 'The CAPM: theory and evidence', *Journal of Economic Perspectives* 18: 25–46.

Feminist Economics (2003) 'Special double issue on "Amartya Sen's Work and Ideas: A Gender Perspective"', *Feminist Economics* 9: 1–342.

Foster, J and Metcalfe, J. (2001) 'Modern evolutionary economic perspectives', in J. Foster and J. Metcalfe (eds) *Frontiers of Evolutionary Economics*, Cheltenham: Edward Elgar.

Fothergill, S. and Gudgin, G. (1985) 'Ideology and methods in industrial location research', in D. Massey and R. Meegan (eds) *Politics and Method*, London: Methuen.

Foucault, M. (1978) *The History of Sexuality* (Vol. 1), New York: Vintage.

Freeman, C. (2000) *High Tech and High Heels in the Global Economy*, London: Duke University Press.

Freire, P. (1972) *Pedagogy of the Oppressed*, London: Penguin.

Froebel, F., Heinrichs, J. and Kreye, O. (1980) *The New International Division of Labour*, Cambridge: Cambridge University Press.

Fukuyama, F. (1992) *The End of History and the Last Man*, Harmondsworth: Penguin.

Gahegan, M. (2000) 'The case for inductive and visual techniques in the analysis of spatial data', *Journal of Geographical Systems* 2: 77–83.

Gamble, D. P. (1949) *Contributions to a Socio-Economic Survey of The Gambia*, London: Colonial Office.

Gamble, D. P. (1955) *Economic Conditions in Two Mandinka Villages*, London: Colonial Office.

Garcia-Ramon, M.-D. (2004) 'The spaces of critical geography: an introduction', *Geoforum* 35: 523–4.

Gardiner, J. (1997) *Gender, Care and Economics*, London: Macmillan.

Geertz, C. (1973) *The Interpretation of Cultures*, New York: Basic Books.

Geiger, S. (1997a) 'Exploring feminist epistemologies and methodologies through the life-histories of TANU women', *Feminist Studies Colloquium*, University of Minnesota.

Geiger, S. (1997b) *TANU Women, Gender and Culture in the Making of Tanganyikan Nationalism, 1955–1965*, Portsmouth, NH: Hienneman.

Geoforum (1999) 'Networks, cultures and elite research: the economic geographer as situated researcher', *Geoforum* 30: 299–363.

Gershuny, J. and Robinson, J. P. (1988) 'Historical shifts in the household division of labor', *Demography* 25: 537–53.

Gertler, M. S. (2004) *Manufacturing Culture: The Institutional Geography of Industrial Practice*, Oxford: Oxford University Press.

Gibson-Graham, J. K. (1994) '"Stuffed if I know!": reflections on post-modern feminist social research', *Gender, Place and Culture* 1: 205–25.

Gibson-Graham, J. K. (1996) *The End of Capitalism as We Knew It*, Oxford: Blackwell.

Gibson-Graham, J. K. (2003) 'Enabling ethical economies: cooperativism and class', *Critical Sociology* 29(2): 1–39.

Gibson-Graham, J. K. (2006) *A Post-capitalist Politics*, Minneapolis: University of Minnesota Press.

Gibson-Graham, J. K. n.d. 'A diverse economy: rethinking economy and economic representation', www.communityeconomies.org.

Gidwani, V. K. (2000) 'The quest for distinction: a re-appraisal of the rural labor process in Kheda district (Gujarat), India', *Economic Geography* 76: 145–68.

Gille, Z. and Ó Riain, S. (2002) 'Global ethnography', *Annual Review of Sociology* 28: 271–95.

Gilmore, R. W. (2005) 'Scholar activists in the mix', *Progress in Human Geography* 29: 177–82.

Gilroy, P. (1993) *The Black Atlantic*, Cambridge, MA: Harvard University Press.

Glasmeier, A. (2005) *An Atlas of Poverty in America: A 40 Year History*, London: Routledge.

Glasmeier, A. and Conroy, M. (1994) 'Global squeeze on rural America', *Institute for Policy Research and Evaluation*, Philadelphia: The Pennsylvania State University.

Glasmeier, A., Gurwitt, R., Kays, A. and Thompson, J. (1995) *Branch Plants and Rural Development in the Age of Globalization*, Washington, DC: The Aspen Institute.

Glassman, J. (1996) 'Whose environmental problems do "we" have?,' MacArthur Consortium Working Paper Series, Institute of International Studies, University of Minnesota.

Glassman, J. (1999) 'State power beyond the "territorial trap"', *Political Geography* 18: 669–96.

Glassman, J. (2002) 'From Seattle (and Ubon) to Bangkok: the scales of resistance to corporate globalization', *Environment & Planning D: Society & Space* 20: 513–33.

Glassman, J. (2004) *Thailand at the Margins*, Oxford: Oxford University Press.

Glassman, J. and Samatar, A. (1997) 'Development geography and the third-world state', *Progress in Human Geography* 21: 164–98.

GLC [Greater London Council] (1985) *London Industrial Strategy*, London: Greater London Council.

Goffman, E. (1959) *The Presentation of Self in Everyday Life*, Garden City, NY: Doubleday.

Gorz, A. (1982) *Farewell to the Working Class*, London: Pluto.

Gowan, P. (1995) 'Neo-Liberal Theory and Practice for Eastern Europe', *New Left Review*, 213: 3–60.

Grabher, G. (2006) 'Trading routes, bypasses, and risky intersections: mapping the travels of "networks" between economic sociology and economic geography', *Progress in Human Geography* 30: 1–27.

Grabher, G. and Hassink, R. (2003) 'Fuzzy concepts, scanty evidence, policy distance? Debating Ann Markusen's assessment of critical regional studies', *Regional Studies* 37: 699–700.

Granovetter, M. (1985) 'Economic action, and social structure: the problem of embeddedness', *American Journal of Sociology* 91: 481–510.

Gregory, D. (2005) 'Geographies, publics and politics', *Progress in Human Geography* 29: 182–93.

Gupta, A. (1995) 'Blurred boundaries: the discourse of corruption, the culture of politics, and the imagined state', *American Ethnologist* 22: 375–402.

Gupta, A. and Ferguson, J. (eds) (1997) *Anthropological Locations*, Berkeley: University of California Press.

Guyer, J. (1981) 'Household and community in African Studies', *African Studies Review* 24: 87–137.

Hamilton, G. G. (1999) 'Introduction', in G. G. Hamilton (ed.) *Cosmopolitan Capitalists*, Seattle, WA: University of Washington Press.

Hamnett, C. (2003) 'Contemporary human geography: fiddling while Rome burns', *Geoforum* 34: 1–3.

Hannan, M. and Freeman, J. (1977) 'The population ecology of organizations', *American Journal of Sociology* 82: 929–64.

Hannerz, U. (1998) 'Transnational research', in R. H. Bernard (ed.) *Handbook of Methods in Cultural Anthropology*, New York: Altamira Press.

Hannerz, U. (2003) 'Being there ... and there ... and there! Reflections on multi-site ethnography', *Ethnography* 4: 201–16.

Hanson, S. (2004) 'Economic geography: then, now and the future', *Annual Meeting of the Association of American Geographers*, March 17, Philadelphia.

Hanson, S., and Pratt, G. (1988) 'Reconceptualizing the links between home and work in urban geography', *Economic Geography* 64: 299–321.

Hanson, S. and Pratt, G. (1995) *Gender, Work, and Space*, London: Routledge.

Haraszti, M. (1977) *A Worker in a Worker's State*, New York: Universe Books.

Haraway, D. (1997) *Modest_Witness@Second_Millenium.FemaleMan_Meets_Oncomouse*, New York: Routledge.

Haraway, D. (2003) *The Companion Species Manifesto*, Chicago: Prickly Paradigm Press.

Hardt, M. and Negri, A. (2000) *Empire*, Cambridge, MA: Harvard University Press.

Hart, G. (1995) 'Gender and household dynamics', in M. G. Quibria (ed.) *Critical Issues in Asian Development*, Hong Kong: Oxford University Press.

Harvey, D. (1982) *The Limits to Capital*, Oxford: Blackwell.

Harvey, D. (1987) 'Three myths in search of a reality in urban studies', *Environment & Planning D: Society & Space* 5: 367–76.

Harvey, D. (1996) *Justice, Nature and the Geography of Difference*, Oxford: Blackwell.

Harvey, D. (1998) 'The body as an accumulation strategy', *Environment & Planning D: Society & Space* 16: 401–21.

Harvey, D. (2000) *Spaces of Hope*, Edinburgh: Edinburgh University Press.

Harvey, D. (2001) *Spaces of Capital*, Edinburgh: Edinburgh University Press.

Harvey, D. (2003) *The New Imperialism*, Oxford: Oxford University Press.

Haswell, M. (1963) *The Changing Pattern of Economic Activity in a Gambian Village*, London: HMSO.

Hayter, R. (2003) 'The war in the woods', *Annals, Association of American Geographers* 933: 706–29.

Hecht, S. and Cockburn, A. (1989) *The Fate of the Forest*, New York: Verso.

Hearnshaw, H. M. and Unwin, D. (eds) (1994) *Visualization in Geographical Information Systems*. Chichester, England: John Wiley & Sons.

Heery, E. (1998) The relaunch of the TUC, *British Journal of Industrial Relations* 36: 339–60.

Hendry, D. (1995) *Dynamic Econometrics*, Oxford: Oxford University Press.

Hendry, D. (2000) *Econometrics: Alchemy or Science?*, Oxford: Oxford University Press.

Henry, N. and Massey, D. (1995) 'Competitive time-space in high technology', *Geoforum* 26: 49–64.

Herbert, S. (1997) *Policing Space*, Minneapolis: University of Minnesota Press.

Herbert, S. (2000) 'For ethnography', *Progress in Human Geography* 24: 550–68.

Herod, A. (1999) 'Reflections on interviewing foreign elites', *Geoforum* 304: 313–27.

Hertz, R. (1997) 'Introduction', in R. Hertz (ed.) *Reflexivity and Voice*, London: Sage

Holgate, J. (2004) *Organising Black and Minority Ethnic Workers: Trade Union Strategies for Recruitment and Inclusion*, London: University of London.

hooks, b. (1990) *Yearning: Race, Gender and Cultural Politics*, Toronto: Between the Lines Press.

Hoover, K. (2001) *Causality in Macroeconomics*, Cambridge: Cambridge University Press.

Houston, D. and Pulido, L. (2002) 'Labor, community and memory', *Environment & Planning D: Society & Space* 20: 401–24.

Hudson, R. (1994) 'Institutional change, cultural transformation, and economic regeneration', in A. Amin and N. J. Thrift (eds) *Globalization, Institutions, and Regional Development in Europe*, Oxford: Oxford University Press.

Hughes, A. (2000) 'Retailers, knowledges and changing commodity networks: the case of the cut flower trade', *Geoforum* 31: 175–90.

Humphrey, C. and Mandel, R. (eds) (2002) *Markets and Moralities: ethnographies of postsocialism*. Oxford: Berg.

Hyden, G. (1980) *Beyond Ujamaa in Tanzania*, Berkeley: University of California Press.

Hyndman, J. (2001) 'The field as here and now, not there and then', *The Geographical Review* 91: 262–72.

Jacka, T. (1994) 'Countering voices: an approach to Asian and feminist studies in the 1990s', *Women's Studies International Forum* 17: 663–72.

Jackson, S. (2004) *Professing Performance: Theatre in the Academy from Philology to Performativity*, Cambridge: Cambridge University Press.

Jamoul, L. (2006) *The Art of Politics*, London: Queen Mary, University of London.

Jarvis, H., Pratt, A. and Wu, P. (2001) *The Secret Life of Cities*, London: Prentice Hall.

Jiménez, O. (2005) 'Innovation-oriented environmental regulations: direct versus indirect regulations, an empirical analysis in small- and medium-sized enterprises in Chile', *Environment & Planning A* 37: 723–50.

Johnston, R. J. (2006) 'The politics of changing human geography's agenda', *Transactions, Institute of British Geographers* 31: 286–303.

Johnston, R. J., Hepple, L., Hoare, T., Jones, K. and Plummer, P. (2003) 'Contemporary fiddling in human geography while Rome burns: has quantitative analysis been largely abandoned – and should it be?', *Geoforum* 34: 157–61.

Kahneman, D. and Tversky, A. (1979) 'Prospect theory: an analysis of decision making under risk', *Econometrica* 47: 263–91.

Katz, C. (1992) 'All the world is staged', *Environment & Planning D: Society & Space* 10: 495–510.

Kawash, S. (1996) 'The autobiography of an ex-coloured man', in E. K. Ginsberg (ed.) *Passing and the Fictions of Identity*, Durham, NC: Duke University Press.

Kennedy, P. (2002) 'Sinning in the basement: what are the rules? The ten commandments of econometrics', *Journal of Economic Surveys* 16: 569–89.

Kessinger, T. (1974) *Vilyatpur, 1848–1968: Social and Economic Change in a North Indian Village*, Berkeley: University of California Press.

Kessler-Harris, A. (1982) *Out to Work: A History of Wage Earning Women in the United States*, Oxford: Oxford University Press.

King, G., Keohane, R. and Verba, S. (1994) *Designing Social Inquiry*, Princeton, NJ: Princeton University Press.

Kobayashi, A. (1994) 'Coloring the field: gender, "race" and the politics of fieldwork', *Professional Geographer* 46: 73–80.

Kobayashi, A. (2005) 'Anti-racist feminism in geography', in L. Nelson and J. Seager (eds) *A Companion to Feminist Geography*, Oxford: Blackwell.

Kong, L. (2004) 'Cultural geography: by whom, for whom?', *Journal of Cultural Geography* 2: 147–50.

Kretzmann, J. and McKnight, J. (1993) *Building communities from the inside out*, Northwestern University, IL: The Asset-Based Community Development Institute, Institute for Policy Research.

Kristman, L. N. (1926) *Georicheskii period velikoi russkoi revoliutsii*, 2a ed., Moscu. Un profundo estudio del Communismo de Guerra.

Krugman, P. (1996) *Development, Geography and Economic Theory*, Cambridge, MA: MIT Press.

Kumarappa, J. C. (1931) *A Survey of Matar Taluka in Kheda District*, Ahmedabad: Vidyapeeth.

Kwan, M. P. (1999) 'Gender, the home-work link, and space-time patterns of non employment activities', *Economic Geography*, 75: 370–94.

Kwan, M. P. (2000) 'Gender differences in space-time constraints', *Area* 32: 145–56.

Kwan, M. P. (2002) 'Feminist visualization: Re-envisioning GIS as a method in feminist geographic research', *Annals, Association of American Geographers*, 92: 645–61.

Kwan, M. P. (2004) 'Beyond difference: From canonical geography to hybrid geographies', *Annals of the Association of American Geographers* 94: 756–63.

Kwan, M. P. (2007) 'Affecting geospatial technologies: Toward a feminist politics of emotion', *The Professional Geographer* 59: Forthcoming.

Kwan, M. P. and Knigge, L. (2006) 'Doing qualitative research using GIS: An oxymoronic endeavor?', *Environment and Planning A*, 38: 1999–2002.

Laclau, E. (1990) *New Reflections on the Revolution of Our Time*, London: Verso.

Larner, W. (1995) 'Theorising difference in Aotearoa/New Zealand', *Gender, Place and Culture* 2: 177–90.

Larner, W. (2002) 'Calling capital: call-centre strategies in New Brunswick and New Zealand', *Global Networks* 2: 133–52.

Latour, B. (1993) *We Have Never Been Modern*, Cambridge, MA: Harvard University Press.

Latour, B. (2003) A prologue in form of a dialog between a student and his (somewhat) Socratic professor.

Latour, B. (2004) 'On using ANT for studying information systems: A (somewhat) Socratic dialogue', in C. Avgerou, C. Ciborra and F. F. Land, *The Social Study of Information and Communication Study*, Oxford: Oxford University Press.

Lawson, T. (1997) *Economics and Reality*, London: Routledge.

Lawson, V. (1995) 'The politics of difference', *Professional Geographer* 47: 449–57.

Lawson, V. (2002) 'Arguments within geographies of movement: the theoretical potential of migrants' stories', *Progress in Human Geography* 24: 173–89.

Lee, R. and Wills, J. (eds) (1997) *Geographies of Economies*, London: Arnold.

Lees, L. (2003) 'Urban geography: 'new' urban geography and the ethnographic void', *Progress in Human Geography* 27: 107–13.

Lenin, V. I. (1956 [1899]) *The Development of Capitalism in Russia*, Moscow: Progress Publishers.

Leyshon, A. and Thrift, N. J. (1997) *Money/Space*, London: Routledge.

Limón, J. (1994) *Dancing with the Devil: Society and Cultural Politics in Mexican-American South Texas*, Madison: University of Wisconsin Press.

Littlefield, D. C. (1981) *Rice and Slaves*, Baton Rouge: Louisiana State University Press.

Lo, A. (2004) 'The adaptive markets hypothesis', *Journal of Portfolio Management* 30: 15–29.

Lovering, J. (1989) 'The restructuring debate', in R. Peet and N. J. Thrift (eds) *New Models in Geography* (Vol. 1), London: Unwin Hyman.

Lovering, J. (1990) 'Fordism's unknown successor: a comment on Scott's theory of flexible accumulation', *International Journal of Urban and Regional Research* 14: 159–74.

Low, S. and Lawrence-Zuniga, D. (2003) *The Anthropology of Space and Place: Locating Culture*, Oxford: Blackwell.

MacEachren, A. M., Wachowicz, M., Edsall, R. and Haug, D. (1999) 'Constructing knowledge from multivariate spatiotemporal data: integrating geographical visualization and knowledge discovery in database methods', *International Journal of Geographical Information Science*, 13: 311–34.

Mackintosh, M. (1981) 'Gender and economics', in K. Young, C. Wolkowitz and R. McCullagh (eds) *Of Marriage and the Market*, London: Routledge.

Maddala, G. S. (2001) *Introduction to Econometrics*, New York: Wiley.

Magnus, J. R. and Morgan, M. S. (1999) *Methodology and Tacit Knowledge: Two Experiments in Econometrics*, New York: Wiley.

Maki, U., Marchionni, C., Oinas, P. and Sayer, A. (2004) 'Scientific realism, geography and economics', *Environment & Planning A* 36: 1717–18.

Manent, P. (1998) *The City of Man*, Princeton, NJ: Princeton University Press.

Marcus, G. E. (1992) 'More critically reflexive than thou: the current identity politics of representation', *Environment & Planning D: Society & Space* 10: 489–93.

Marcus, G. E. (1995) 'Ethnography in/of the world system: the emergence of multi-sited ethnography', *Annual Review of Anthropology* 24: 95–117.

Markusen, A. (1999) 'Fuzzy concepts, scanty evidence, policy distance: the case for rigor and policy relevance in critical regional studies', *Regional Studies* 33: 869–84.

Markusen, A. (2001) 'The activist intellectual', *Antipode* 33: 39–48.

Martin, R. and Sunley, P. (2001) 'Rethinking the "economic" in economic geography: broadening our vision or losing our focus?', *Antipode* 33: 148–61.

Martin, R. L. (2001) 'Geography and public policy', *Progress in Human Geography* 25: 189–209.

Marx, K. (1976 [1867]) *Capital* (Vol. 1), Harmondsworth: Penguin.

Massey, D. (1979) 'In what sense a regional problem?', *Regional Studies* 13: 233–43.

Massey, D. (1984) *Spatial Divisions of Labour*, London: Macmillan.

Massey, D. (1992) 'A place called home?', *New Formations* 17: 3–15.

Massey, D. (1995) 'Masculinity, dualisms and high technology', *Transactions, Institute of British Geographers* 20: 487–99.

Massey, D. (2000) 'Practising political relevance', *Transactions, Institute of British Geographers* 25: 131–3.

Massey, D. (2001) 'Geography and the agenda', *Progress in Human Geography* 25: 5–17.

Massey, D. (2005) *For Space*, London: Sage.

Massey, D. and Meegan, R. (1982) *The Anatomy of Job Loss*, London: Methuen.

Massey, D. and Meegan, R. (eds) (1985a) 'Introduction: the debate', in D. Massey and R. Meegan (eds) *Politics and Method*, London: Methuen.

Massey, D. and Meegan, R. (eds) (1985b) *Politics and Method*, London: Methuen.

Massey, D., Quintas, P. and Wield, D. (1992) *High-Tech Fantasies*, London: Routledge.

McAleer, M. (1994) 'Sherlock Holmes and the search for truth: a diagnostic tale', *Journal of Economic Surveys* 8: 317–70.

McDowell, L. (1992) 'Doing gender', *Transactions, Institute of British Geographers* 17: 399–416.

McDowell, L. (1992b) 'Valid games? A response to Erica Schoenberger, *Professional Geographer* 44: 212–15.

McDowell, L. (1997) *Capital Culture*, Oxford: Blackwell.

McDowell, L. (1998) 'Elites in the City of London: some methodological considerations', *Environment & Planning A* 30: 2133–46.

McDowell, L. (2000) 'Economics, geography and gender', in G. L. Clark, M. Feldman and M. Gertler (eds) *The Oxford Handbook of Economic Geography*, Oxford: Oxford University Press.

McDowell, L. (2003) *Redundant masculinities?*, Oxford: Blackwell.

McDowell, L. (2004) 'Masculinity, identity and labour market change', *Geografiska Annaler* 86: 45–56.

McDowell, L. (2005a) 'For love and money: some critical comments on welfare to work policies', *Journal of Economic Geography* 5: 365–79.

McDowell, L. (2005b) *Hard Labour: The Forgotten Voices of Latvian Migrant Workers*, London: UCL Press.

McDowell, L. and Massey, D. (1984) 'A woman's place?', in D. Massey and J. Allen (eds) *Geography Matters!*, Cambridge: Cambridge University Press.

McNamara, R. and VanDeMark, B. (1995) *In Retrospect: The Tragedy and Lessons of Vietnam*, New York: Times Books.

Metcalfe, J. (1994) 'Competition, Fisher's principle and increasing returns in the selection process', *Journal of Evolutionary Economics* 4: 327–46.

Meurs, M. and Ranasinghe, R. (2003) 'De-development in post-socialism: conceptual and measurement issues', *Politics and Society*, 31: 31–54.

Mikhova, D. and Pickles, J. (1994) 'Environmental data and social change in Bulgaria: problems and prospects', *The Professional Geographer*, 46: 229–36.

Milkman, R. (1987) *Gender at Work*, Chicago: University of Chicago Press.

Mirchandani, K. (2004) 'Practices of global capital: gaps, cracks and ironies in transnational call centres in India', *Global Networks* 4: 355–73.

Mitchell, J. C. (1983) 'Case and situational analysis', *Sociological Review* 31: 187–211.

Mitchell, T. (2002) *Rule of Experts*, Berkeley: University of California Press.

Morgan, K. and Sayer, A. (1988) *Microcircuits of Capital*, Cambridge: Polity Press.

Mountz, A. (2003) *Embodied Geographies of the Nation-State*, Vancouver: University of British Columbia Press.

Murdoch, J. (1997) 'Inhuman/nonhuman/human: actor-network theory and the prospects for a nondualistic and symmetrical perspective on nature and society', *Environment & Planning D: Society & Space* 15: 731–56.

Myers, G. and Papageorgiou, Y. (1991) 'Homo economicus in perspective', *The Canadian Geographer* 35: 380–99.

Nagar, R. (1995) *Making and Breaking Boundaries: Identity Politics among South Asian in Postcolonial Dar es Salaam*, Minneapolis: University of Minnesota Press.

Nagar, R. (2002) 'Women's theatre and redefinitions of public, private and politics in north India', *ACME* 1: 55–72.

Nagar, R. (2006) 'Local and global', in S. Aitken and G. Valentine (eds) *Approaches to Human Geography*, London: Sage.

Nancy, J.-L. (1991) 'Of being-in-common', in The Miami Theory Collective (ed.) *Community at Loose Ends*, Minneapolis: University of Minnesota Press.

National Research Council (1996) *Lost Crops of Africa*, Washington, DC: National Academy Press.

Nelson, J. (1992) 'Gender, metaphor and the definition of economics', *Economics and Philosophy* 8: 103–25.

Nelson, L. and Seager, J. (2005) 'Introduction', in L. Nelson and J. Seager (eds) *A Companion to Feminist Geography*, London: Blackwell.

Nelson, R. (1995) 'Recent evolutionary thinking about economic growth', *Journal of Economic Literature* 33: 48–90.

Nelson, R. and Winter, S. (1982) *An Evolutionary Theory of Economic Growth*, Cambridge: Cambridge University Press.

Nussbaum, M. C. (1999) *Sex and Justice*, Oxford: Oxford University Press.

Nussbaum, M. C. (2000) *Women and Human Development*, Cambridge: Cambridge University Press.

O'Connor, A. (2002) *Poverty Knowledge: Social Science, Social Policy and the Poor in Twentieth Century US History*, Princeton, NJ: Princeton University Press.

O'Neill, J. (1994) 'Essentialism and the market', *The Philosophical Forum* XXVI: 87–100.

O'Neill, J. (2004) 'Commerce and the language of value', in M. S. Archer and W. W. Outwaite (eds) *Defending Objectivity: Essays in Honour of Andrew Collier*, London: Routledge.

Olds, K. (2001) *Globalization and Urban Change*, Oxford: Oxford University Press.

Olds, K. (2005) 'Articulating agendas and traveling principles in the layering of new strands of academic freedom in contemporary Singapore', in B. Czarniawska and G. Sevón (eds) *Where Translation is a Vehicle, Imitation its Motor, and Fashion Sits at the Wheel*, Malmö: Liber AB.

Olds, K. and Thrift, N. J. (2005) 'Cultures on the brink: reengineering the soul of capitalism – on a global scale', in A. Ong and S. Collier (eds) *Global Assemblages*, Oxford: Blackwell.

Olds, K. and Yeung, H. (1999) 'Reshaping Chinese business networks in a globalising era', *Environment & Planning D: Society & Space* 17: 535–55.

Ong, A. (1987) *Spirits of Resistance and Capitalist Discipline*, Albany, NY: SUNY Press.

Ong, A. (1995) 'Women out of China', in R. Behar and D. Gordon (eds) *Women Writing Culture*, Berkeley: University of California Press.

Ong, A. (1999) *Flexible Citizenship*, Durham, NC: Duke University Press.

Openshaw, S. (1997) 'Towards a computationally minded scientific human geography', *Environment & Planning A* 30: 317–32.

Orrù, M., Biggart, N. and Hamilton, G. (1997) *The Economic Organization of East Asian Capitalism*, London: Sage.

Ortner, S. (1996) *Making Gender: The Politics and Erotics of Culture*, Boston: Beacon Press.

Orwell, G. (2004) 'Why I write', Harmondsworth: Penguin.

Osterweil, M. (2004) 'A cultural-political approach to reinventing the political', *International Social Science Journal* 18: 495–506.

Parsons, J. J. (1972) 'Spread of African pasture grasses to the American tropics', *Journal of Range Management* 25: 12–17.

Paskaleva, K., Shapira, P., Pickles, J. and Koulov, B. (eds) (1998) *Bulgaria in Transition*, New York: Ashgate.

Patai, D. (1991) 'US academics and Third World women: is ethical research possible?', in S. B. Gluck and D. Patai (eds) *Women's Words*, New York: Routledge.

Pateman, C. (1988) *The Sexual Contract*, Cambridge: Polity Press.

Pavlínek, P. and Pickles, J. (2000) *Environmental Transitions: Transformation and Ecological Defense in Central and Eastern Europe*, London: Routledge.

Peck, J. (1999) 'Grey geography?' *Transactions, Institute of British Geographers* 24: 131–35.

Peck, J. (2000) 'Jumping in, joining up, getting on', *Transactions, Institute of British Geographers* 25: 255–8.

Peck, J. (2002) 'Labor, zapped/growth, restored? Three moments of neoliberal restructuring in the American labor market', *Journal of Economic Geography* 2: 179–220.

Peck, J. (2005) 'Economic sociologies in space', *Economic Geography* 82: 120–75.

Peet, R. and Watts, M. (eds) (1996) *Liberation Ecologies*, New York: Routledge.

Peluso, N. L. (1992) *Rich Forests, Poor People*, Berkeley: University of California Press.

Persky, J. (1995) 'The etyology of Homo Economicus', *Journal of Economic Perspectives* 9: 221–31.

Pickles, J. (1995) 'Restructuring state enterprises: industrial geography and eastern european transitions', *Geographische Zeitschrift*, 83: 114–31.

Pickles, J. (2001) *There are No Turks in Bulgaria*, Max Planck Research Institute for Social Anthropology, Working Paper, Halle/Salle.

Pickles, J. (2004) 'Disseminated Economies and the New Economic Geographies of Post-Socialism'. Invited paper, 'Rethinking the Economic Geographies of Post-Socialism' workshop, Queen Mary College, University of London, September 24.

Pickles, J. (2005) '"New cartographies" and the decolonisation of European geographies', *Area* 35: 355–64.

Pickles, J., Nikolova, M., Staddon, C., Velev, S., Mateeva, Z. and Popov, A. (2002) 'Bulgaria', in D. Turnock and F. Carter (eds), *Environmental Problems in East-Central Europe*, London: Routledge.

Pickles, J. and Smith, A. (1998) *Theorizing Transition: The Political Economy of Postcommunist Tranformations*, London: Routledge.

Pickles, J. and Smith, A. (2005) 'Technologies of Transition: Foreign Investment and the (Re-)Articulation of East Central Europe into the Global Economy', in D. Turnock (ed.) *Foreign Direct Investment and Regional Development in East Central Europe and the Former Soviet Union*, London: Ashgate.

Pickles, J., Smith, A., Bucek, M., Begg, R. and Roukova, P. (2006) 'Upgrading and diversification in the East European apparel industry', *Environment and Planning A* 38: 2305–24.

Pinder, D. (2002) 'In defence of utopian urbanism: imagining cities after the "end of utopia"', *Geografiska Annaler* 84B: 229–41.

Pinder, D. (2005) *Visions of the City: Utopianism, Power and Politics in Twentieth-Century Urbanism*, Edinburgh: Edinburgh University Press.

Plummer, P. and Sheppard, E. (2001) 'Must emancipatory geography be qualitative: a response to Amin and Thrift', *Antipode* 33: 194–9.

Plummer, P. and Sheppard, E. (2006) 'Geography matters: agency, structure, and dynamics at the intersection of economics and geography', *Journal of Economic Geography* 6: 619–37.

Plummer, P. and Taylor, M. (2001a) 'Theories of local economic growth: concepts, models and measurement', *Environment & Planning A* 33: 385–99.

Plummer, P. and Taylor, M. (2001b) 'Theories of local economic growth: model specification and empirical validation', *Environment & Planning A* 33: 219–36.

Plummer, P. and Taylor, M. (2003) 'Theory and praxis in economic geography: "enterprising" and local growth in a global economy', *Environment & Planning C* 21: 633–49.

Polanyi, K. (1944) *The Great Transformation*, New York: Rinehart & Company.

Pollard, J., Henry, N., Bryson, J. and Daniels, P. (2000) 'Shades of grey? Geographers and policy', *Transactions, Institute of British Geographers* 25: 243–8.

Poole, R. (1991) *Morality and Modernity*, London: Routledge.

Portes, A. (2001) 'Introduction: the debates and significance of immigrant transnationalism', *Global Networks* 1: 181–93.

Poulantzas, N. (1975) *Classes in Contemporary Capitalism*, London: New Left Books.

Power, M. (1997) *The Audit Society*, Oxford: Oxford University Press.

Pratt, G. (2000) 'Research performances', *Environment & Planning D, Society & Space* 18: 639–51.

Pratt, G. (2001) 'Studying immigrants in focus groups', in P. Moss (ed.) *Geography and Gender: A Guide to Methodology*, Oxford: Blackwell.

Pratt, G. (2004) *Working Feminism*, Edinburgh: Edinburgh University Press.

Pratt, G. and Kirby, E. (2003) 'Performing nursing: BC Nurses' Union theatre project. ACME', 2: 14–32.

Proctor, J. and Smith, D. S. (eds) (1999) *Geography and Ethics*, London: Routledge.

Purvis, T. and Hunt, A. (1993) 'Discourse, ideology, discourse, ideology, discourse, ideology', *British Journal of Sociology* 44: 474–99.

Rabinow, P. (2003) *Anthropos Today: Reflections on Modern Equipment*, Princeton, NJ: Princeton University Press.

Raffles, H. (2002) *In Amazonia*, Princeton, NJ: Princeton University Press.

Rigby, D. and Essletzbichler, J. (1997) 'Evolution, process variety, and regional trajectories of technological change in US manufacturing', *Economic Geography* 73: 269–84.

Rigby, D. and Essletzbichler, J. (2006) 'Technological variety, technological change and a geography of production techniques', *Journal of Economic Geography* 6: 71–89.

Robbins, P. (2001) 'Fixed categories in a portable landscape', *Environment & Planning A* 33: 161–79.

Robbins, P. (2004) 'Comparing invasive networks: the cultural and political biographies of invasion', *Geographical Review* 94: 139–56.

Robbins, P. (2004a) *Political Ecology*, Oxford: Blackwell.

Robbins, P. and Luginbuhl, A. (2005) 'The last enclosure: resisting privatization of wildlife in the Western United States', *Capitalism, Nature, Socialism* 16: 45–61.

Robbins, P., Polderman, A.-M. and Birkenholtz, T. (2001) 'Lawns and toxins: an ecology of the city', *Cities* 18: 369–80.

Robbins, P. and Sharp, J. (2003) 'Producing and consuming chemicals: the moral economy of the American lawn', *Economic Geography* 79: 425–51.

Rojanaphruk, P. (2005) 'Four years of disappointment under Thaksin', *The Nation* (Bangkok). 25 May, p. 5A.

Rosaldo, M. (1980) 'The use and abuse of anthropology: reflections on feminist and cross-cultural understanding', *Signs* 5: 389–416.

Rose, G. (1997) 'Situating knowledges', *Progress in Human Geography* 21: 305–20.

Said, E. W. (2001) 'The public role of writers and intellectuals', *Nation* 17 September: 27–34.

Salman, M. D. (2003) 'Chronic wasting disease in deer and elk: scientific facts and findings', *Journal of Veterinary Medical Science* 65: 761–8.

Salzinger, L. (2003) *Genders in Production: Making Workers in Mexico's Global Factories*, Berkeley: University of California Press.

Sangtin Writers and Nagar, R. (2006) *Playing with Fire: Feminist Thought and Activism Through Seven Lives in India*, New Delhi: Zubaan.

Sassen, S. (2001) *The Global City* (2nd edn), Princeton, NJ: Princeton University Press.

Sauer, C. O. (1965) 'The morphology of landscape', in J. Leighly (ed.) *Land and Life*, Berkeley: University of California Press.

Sauer, C. O. (1966) *The Early Spanish Main*, Berkeley: University of California Press.

Savage, M., Barlow, J., Duncan, S. and Saunders, P. (1987) "Locality research": the Sussex programme on economic restructuring, social change and the locality', *Quarterly Journal of Social Affairs* 3: 27–51.

Saviotti, P. and Metcalfe, S. (1991) *Evolutionary Theories of Economic and Technological Change*, Chur, Switzerland: Harwood.

Saxenian, A. (1994) *Regional Advantage*, Cambridge, MA: Harvard University Press.

Sayer, A. (1984) *Method in Social Science*, London: Hutchinson.

Sayer, A. (1985) 'Industry and space: a sympathetic critique of radical research', *Environment & Planning D: Society & Space* 3: 3–29.

Sayer, A. (1987) 'Hard work and its alternatives', *Environment & Planning D: Society & Space* 5: 395–9.

Sayer, A. (1995) *Radical Political Economy: A Critique*, Oxford: Blackwell.

Sayer, A. (2000a) 'Moral economy and political economy', *Studies in Political Economy* 61: 79–103.

Sayer, A. (2000b) *Realism and Social Science*, London: Sage.

Sayer, A. (2004) 'Restoring the moral dimension in social scientific accounts', in M. S. Archer and W. Outhwaite (eds) *Defending Objectivity: Essays in Honour of Andrew Collier*, London: Routledge.

Sayer, A. (2005) *The Moral Significance of Class*, Cambridge: Cambridge University Press.

Sayer, A. and Morgan, K. (1985) 'A modern industry in a declining region: links between method, theory and policy', in D. Massey and R. Meegan (eds) *Politics and Method*, London: Methuen.

Sayer, A. and Storper, M. (1997) 'Ethics unbound: for a normative turn in social theory', *Environment & Planning D: Society & Space* 15: 1–17.

Schauber, E. M. and Woolf, A. (2003) 'Chronic wasting disease in deer and elk: a critique of current models and their application', *Wildlife Society Bulletin* 31: 610–16.

Schlosser, E. (2001) *Fast Food Nation*, New York: Houghton Mifflin.

Schoenberger, E. (1991) 'The corporate interview as a research method in economic geography', *Professional Geographer* 43: 180–9.

Schoenberger, E. (1992) 'Self-criticism and self-awareness in research: a reply to Linda McDowell', *Professional Geographer* 44: 215–18.

Schoenberger, E. (1997) *The Cultural Crisis of the Firm*, Oxford: Blackwell.

Schoenberger, E. (2001) 'Corporate autobiographies: the narrative strategies of corporate strategists', *Journal of Economic Geography* 1: 277–98.

Scott, A. J. (1983) 'Industrial organization and the logic of intra-metropolitan location I. theoretical considerations', *Economic Geography* 59: 233–50.

Scott, A. J. (1988a) 'Flexible production systems and regional development: the rise of new industrial spaces in North America and Western Europe', *International Journal of Urban and Regional Research* 12: 171–85.

Scott, A. J. (1988b) *Metropolis: From the Division of Labor to Urban Form*, Berkeley: University of California Press.

Scott, A. J. (1991) 'Flexible production systems: analytical tasks and theoretical horizons – a reply to Lovering', *International Journal of Urban and Regional Research* 15: 130–34.

Scott, A. J. (2000) 'Economic geography: the great half-century', in G. L. Clark, M. Feldman and M. S. Gertler (eds) *The Oxford Handbook of Economic Geography*, Oxford: Oxford University Press.

Scott, J. (1998) *Seeing Like a State*, New Haven, CT: Yale University Press.

Scott, J. C. (1976) *The Moral Economy of the Peasant*, New Haven, CT: Yale University Press.

Sen, A. (1981) *Poverty and Famine*, Oxford: Clarendon Press.

Sen, A. (1999) *Development as Freedom*, Oxford: Oxford University Press.

Sevenhuijsen, S. (1998) *Citizenship and the Ethic of Care*, London: Routledge.

Shanin, T. (1972) *The Awkward Class: Political Sociology of the Peasantry in a Developing Society: Russia, 1910–1925*, Oxford: Clarendon Press.

Shaw, M. and Munro, I. (2001) 'Valley of despair' *The Age* 23 October, p. 1.

Sheehan, N. (1988) *A Bright Shining Lie*, New York: Random House.

Sheppard, E. (1984) 'Value and exploitation in a capitalist space economy', *International Regional Science Review* 9: 97–107.

Sheppard, E. (2001) 'Quantitative geography: Representations, practices and possibilities', *Environment and Planning D: Society & Space* 19: 535–54.

Sheppard, E. (2002) 'The spaces and times of globalization: places, scale, networks, and positionality', *Economic Geography* 78: 307–30.

Sheppard, E. and Barnes, T. J. (2000) *The Companion to Economic Geography*, Oxford: Blackwell.

Sheppard, E. and Barnes, T. (1990) *The Capitalist Space Economy*, London: Unwin-Hyman.

Shohat, E. (1996) 'Notes on the postcolonial', in P. Mongia (ed.) *Contemporary Postcolonial Theory: A Reader*, London: Arnold.

Singer, L. (1991) 'Recalling a community at loose ends', in The Miami Theory Collective (ed.) *Community at Loose Ends*, Minneapolis: University of Minnesota Press.

Singh, R. and Nagar, R. (2006) 'In the aftermath of critique: the journey after Sangtin Yatra', in S. Raju, S. Kumar and S. Corbridge (eds) *Colonial and Postcolonial Geographies of India*, London: Sage.

Smith, A. (1998) *Reconstructing the Regional Economy: Industrial Transformation and Regional Development in Slovakia*, Cheltenham: Edward Elgar.

Smith, A. (2000) 'Employment restructuring and household survival in "postcommunist transition"', *Environment and Planning A*, 32: 10, 1759–80.

Smith, A. (2002) 'Culture/economy and spaces of economic practice: positioning households in post-communism', *Transactions, Institute of British Geographers*, 27: 2, 232–50.

Smith, A. (2007) 'Articulating neo-liberalism: diverse economies and everyday life in post-socialist cities', in H. Leitner, J. Peck and E. Sheppard (eds) *Contesting neoliberalism: urban frontiers*, New York: Guilford.

Smith, A. and Hardy, J. (2004) 'Governing regions, governing transition: firms, institutions and regional change in East-Central Europe', in A. Wood and D. Valler (eds) *Placing Institutions,* Aldershot: Ashgate.

Smith, A. and Stenning, A. (2006) 'Beyond household economies: articulations and spaces of economic practice in post-socialism', *Progress in Human Geography* 30: 190–213.

Smith, D. S. (2000) *Moral Geographies*, Edinburgh: Edinburgh University Press.

Smith, N. (1987) 'Dangers of the empirical turn: some comments on the CURS initiative', *Antipode* 19: 59–68.

Soja, E. (1989) *Postmodern Geographies*, London: Verso.

Soper, K. (1995) 'Forget Foucault?, *New Formations* 25: 21–7.

Spinosa, C., Flores, F. and Dreyfus, H. L. (1997) *Disclosing New Worlds: Entrepreneurship, Democratic Action and the Cultivation of Solidarity*, Cambridge, MA: MIT Press.

Spivak, G. C. (1986) 'Scattered speculations on the question of value', in G. C. Spivak (ed.) *In Other Worlds*, New York: Routledge.

Standing, G. (2002) 'The babble of euphemisms: re-embedding social protection in "transformed" labour markets', in A. Rainnie A. Smith and A. Swain (eds) *Work, Employment and Transition*. London: Routledge.

Steedman, I. (1977) *Marx after Sraffa*, London: New Left Books.

Storper, M. (1987) 'The new industrial geography, 1985–1986', *Urban Geography* 8: 585–98.

Storper, M. (1992) 'The limits to globalization: technology districts and international trade', *Economic Geography* 68: 60–93.

Storper, M. and Salais, R. (1997) *Worlds of Production*, Cambridge, MA: Harvard University Press.

Storper, M. and Scott, A. J. (1986) 'Production, work, territory: contemporary realities and theoretical risks', in A. J. Scott and M. Storper (eds) *Production, Work, Territory*, Boston: Allen and Unwin.

Storper, M. and Walker, R. (1989) *The Capitalist Imperative*, Oxford: Blackwell.

Strange, S. (1997) 'The future of global capitalism; or will divergence persist forever?', in C. Crouch and W. Streeck (eds) *Political Economy of Modern Capitalism: Mapping Convergence and Diversity*, London: Sage.

Strathern, M. (2004) 'Laudable aims and problematic consequences, or, the "flow" of knowledge is not neutral', *Economy and Society* 33. 550–61.

Strathern, M. (2006) 'Bulletproofing: A tale from the United Kingdom', in A. Riles (ed.) *Documents: Artifacts of Modern Knowledge*, Ann Arbor: University of Michigan Press.

Sylvester, C. (ed.) (1998) *The Penguin Book of Columnists*, London: Penguin.

Taylor, P. (1990) 'Editorial comment: GKS', *Political Geography Quarterly* 9: 211–12.

Tejapira, K. (2001) *Commodifying Marxism: The Formation of Modern Thai Radical Culture. 1927–1958*, Kyoto: Kyoto University Press.

Thrift, N. (2000) 'Pandora's box? Cultural geographies of economics', in G. L. Clark, M. P. Feldman, and M. S. Gertler (eds) *The Oxford Handbook of Economic Geography*, Oxford: Oxford University Press.

Thrift, N. J. (1997) 'The still point: resistance, expressive embodiment and dance', in S. Pile and M. Keith (eds) *Geographies of Resistance*, New York: Routledge.

Thrift, N. J. (2000) 'Afterwords', *Environment & Planning D: Society & Space* 18: 213–55.

Thrift, N. J. (2005) *Knowing Capitalism*, London: Sage.

Thrift, N. J. and Olds, K. (1996) 'Refiguring the economic in economic geography', *Progress in Human Geography* 203: 311–37.

Tickell, A. (1995) 'Reflections on "Activism and the academy"', *Environment & Planning D: Society & Space* 13: 235–7.

Tickell, A. and Peck, J. (1992) 'Accumulation, regulation and the geographies of post-Fordism', *Progress in Human Geography* 16: 190–218.

Tilly, C. (1997) 'Invisible elbow', *Sociological Forum* 11: 589–601.

Tippett, G. (1997) 'Valley of the dole' *The Sunday Age*, 26 January, p. 5.

Tivers, J. (1985) *Women attached: The daily lives of women with young children*, London: Croom Helm.

Treanor, J. (2003) 'Furse hits back at "frivolous" media' *The Guardian* 27 February, p. 26.

Tronto, J. (1994) *Moral Boundaries*, London: Routledge.

Tsang, E. W. K. (2002) 'Learning from overseas venturing experience – the case of Chinese family businesses', *Journal of Business Venturing* 17: 21–40.

Tsui-Auch, L. S. (2004) 'The professionally managed family-ruled enterprise: ethnic Chinese business in Singapore', *Journal of Management Studies* 41: 693–723.

Turner, B. L., Villar, S. C., Foster, D., Geoghegan, J., Keys, E., Klepeis, P., Lawrence, D., Mendoza, P. M., Manson, S., Ogneva-Himmelberger, Y., Plotkin, A. B., Salicrup, D. P. Chowdhury, R. R., Savitsky, B., Schneider, L. and Schmook, B. (2001) 'Deforestation in the Southern Yucatan peninsular region: an integrative approach', *Forest Ecology and Management* 154: 353–70.

Vaddhanaphuti, C. (1984) *Cultural and Ideological Reproduction in Rural Northern Thai Society*, Standford, CA: Stanford University Press.

van Duijn, J. (1983) *The Long Wave in Economic Life*, London: Allen and Unwin.

Varela, F. (1992) *Ethical Know-how: Action, Wisdom and Cognition*, Stanford, CA: Stanford University Press.

Vaughan, G. (1997) *For-giving: a feminist critique of exchange*. Austin, TX: Plain View Press.

Visweswaran, K. (1994) *Fictions of Feminist Ethnography*, Minneapolis: University of Minnesota Press.

Voeks, R. (1997) *Sacred Leaves of Candomblé*, Austin: University of Texas Press.

Wade, R. (1990) *Governing the Market*, Princeton, NJ: Princeton University Press.

Wajcman, J. (2004) *TechnoFeminism*, Cambridge: Polity Press.

Walker, R. (1985) 'Is there a service economy? The changing capitalist division of labor', *Science and Society* 49: 42–83.

Ward, K. G. and Jones, M. (1999) 'Researching local elites', *Geoforum* 30: 301–12.

Watts, D. (1987) *The West Indies: Patterns of Development, Culture and Environmental Change since 1492*, Cambridge University Press.

Watts, M. (1983) *Silent Violence: Food, Famine and Peasantry in Northern Nigeria*, Berkeley: University of California Press.

Webber, M. (1982) 'Agglomeration and the regional question', *Antipode* 14: 1–11.

Webber, M. (1987) 'Quantitative measurement of some Marxist categories', *Environment & Planning A* 19: 1302–22.

Webber, M. and Rigby, D. (1996) *The Golden Age Illusion*, New York: Guilford Press.

Weil, P. (1982) 'Agrarian production, intensification and underdevelopment: Mandinka women of The Gambia in time perspective', in C. Curtis (ed.) *Proceedings of the Title XII Conference on Women in Development*, Newark: University of Delaware.

Wheatley, N. (2001) *The Life and Myth of Charmian Clift*, Pymble, NSW: Flamingo.

Whitley, R. (1999) *Divergent Capitalisms*, New York: Oxford University Press.

Wills, J. (2002) *Union Futures: Building Networked Trade Unionism in the UK*, London: Fabian Society.

Wills, J. (2005) 'The geography of union organising in low-paid service industries in the UK', *Antipode* 37: 139–59.

Wills, J. with Hurley, J. (2005) 'Action research: tracing the threads of labour in the global garment industry', in A. Hale and J. Wills (eds) *Threads of Labour: Garment Industry Supply Chains from the Workers' Perspective*, Oxford: Blackwell.

Winichakul, T. (1994) *Siam Mapped: A History of the Geo-body of a Nation*, Honolulu: University of Hawai'i Press.

Wold, R. (2003) 'Charade '04: the triumph of the image', *Columbia Journalism Review* 42: 22–6.

Wolf, D. L. (1992). *Factory Daughters: Gender, Household Dynamics, and Rural Industrialization in Java*, Berkeley: University of California Press.

Wolf, D. L. (1997) 'Situating feminist dilemmas in fieldwork', in D. Wolf (ed.) *Feminist Dilemmas in Fieldwork*, Boulder, CO: Westview Press.

Wolf, E. (1966) *Peasants*, Englewood Cliffs, NJ: Prentice-Hall.

Wood, P. (1974) *Black Majority*, New York: Knopf.

Woolf, V. (1938) 'The modern essay', in V. Woolf (ed.) *The Common Reader*, Harmondsworth: Penguin.

World Bank (1993) *The East Asian Miracle: Economic Growth and Public Policy*, New York: World Bank.

World Bank (1996) *From Plan to Market: World Development Report*, Washington DC: World Bank.

World Commission on Dams (WCD) (2000) *Pak Mun Case Study: Final Draft Report*, London: Earthscan Press.

Wright, M. (1997) 'Crossing the factory frontier', *Antipode* 29: 278–302.

Yao, S. (2002) *Confucian Capitalism*, London: RoutledgeCurzon.

Yapa, L. (2002) 'How the discipline of geography exacerbates poverty in the Third World', *Futures* 34: 33–46.

Yeung, H. (1997) 'Business networks and transnational corporations', *Economic Geography* 73: 1–25.

Yeung, H. (1998) *Transnational Corporations and Business Networks*, London: Routledge.

Yeung, H. (2002) *Entrepreneurship and the Internationalisation of Asian Firms*, Cheltenham: Edward Elgar.

Yeung, H. (2003) 'Practicing new economic geographies', *Annals, Association of American Geographers* 93: 442–62.

Yeung, H. (2004) *Chinese Capitalism in a Global Era*, London: Routledge.

Yeung, H. and Lin, G. C. S. (2003) 'Theorizing economic geographies of Asia', *Economic Geography* 79: 107–28.

Yin, R. K. (2003) *Case Study Research*, London: Sage.

Zhou, Y. and Tseng, Y.-F. (2001) 'Regrounding the "ungrounded empires"', *Global Networks* 12: 131–54.

Zimmerer, K. and Bassett, T. (eds) (2003) *Political Ecology*, New York: Guilford Press.

Žižek, S. (2001) *Did Someone Say Totalitarianism? Four Interventions in the Misuse of a Notion*, London: Verso Books.

Index